Vanessa Dummer Marques

Maize Water Requirements in São Miguel do Oeste/SC via ISAREG

Vanessa Dummer Marques

Maize Water Requirements in São Miguel do Oeste/SC via ISAREG

Observed and Future

ScienciaScripts

Imprint

Any brand names and product names mentioned in this book are subject to trademark, brand or patent protection and are trademarks or registered trademarks of their respective holders. The use of brand names, product names, common names, trade names, product descriptions etc. even without a particular marking in this work is in no way to be construed to mean that such names may be regarded as unrestricted in respect of trademark and brand protection legislation and could thus be used by anyone.

Cover image: www.ingimage.com

This book is a translation from the original published under ISBN 978-613-9-62886-5.

Publisher:
Sciencia Scripts
is a trademark of
Dodo Books Indian Ocean Ltd. and OmniScriptum S.R.L publishing group

120 High Road, East Finchley, London, N2 9ED, United Kingdom
Str. Armeneasca 28/1, office 1, Chisinau MD-2012, Republic of Moldova, Europe
Printed at: see last page
ISBN: 978-620-7-75477-9

SUMMARY

1. INTRODUCTION

Water is one of the most important resources on Planet Earth and essential to life, being the most precious resource available to humanity. Even so, it is possible to see the complete neglect and misuse of this resource in various activities, whether by the population in general, through excessive consumption, through contamination and neglect of water that is still pure and potable due to urban sewage discharged untreated into watercourses, or even in industry and agriculture, through the inappropriate use of chemical additives and agrochemicals in crops, which can contaminate water sources or the water table, potentially causing an irreversible environmental impact.

According to the Human Development Report (UNDP, 2006), it is estimated that by the middle of 2025 water scarcity will be the main cause of conflict between nations, because although around 70.8% of the earth's surface is occupied by water, most of it is concentrated in the oceans. Of this, only 2.2% is fresh water (ANA, 2007), which corresponds to around 10.5 million km^3 of water, distributed between groundwater, rivers, lakes, swamps, soil moisture and vapor in the atmosphere. Of these 10.5 million km3 of water, approximately 98.7% corresponds to groundwater and only 1.3% corresponds to the volume of fresh surface water (rivers and lakes) directly available for human consumption. This volume is enough to meet 6 to 7 times the annual minimum that each inhabitant of the planet needs, considering the current population of around 7 billion (UN, 2013).

Even in the face of such water availability, concerns about excessive water consumption in agriculture have been pointed out as one of the reasons for the future planetary crisis. In Brazil, it is estimated that agriculture is the activity responsible for the greatest demand for water in the country, where irrigation consumes an average of 65% of the country's fresh water (Embrapa, 2009). According to data from the National Water Agency (ANA, 2007), Brazil cultivates approximately 60 million hectares of land and of these, around 6% (3.6 million) are irrigated, with growth expected to rise to a further 30 million in the coming years. However, around 93% of the 3.6 million hectares irrigated use the least efficient methods, such as surface irrigation, which accounts for 56% of the total irrigated area, followed by sprinkler irrigation, where the central pivot system (19%) and conventional sprinkler irrigation (18%) predominate.

Data from the FAO (Food and Agriculture Organization) indicate that around 60% of the water used in irrigation projects around the world is lost through evaporation or underground percolation. In turn, the United Nations (UN, 2003) estimates that a 10% reduction in the water used in agriculture would be enough to supply twice the world's population.

According to a joint report by the UN/FAO and the Organization for Economic Cooperation and Development (OECD), Brazil's agricultural production is expected to register the highest growth in the world, of more than 40% by 2019 compared to the period between 2007 and 2009, which represents a worrying index related to the future availability of water resources (Paz et al., 2000).

Thus, the growth in demand for food to meet the needs of the population is forcing an increase in agricultural production, making it necessary to develop techniques that maximize productivity by combining them with techniques aimed at the rational consumption of water resources for agriculture, focusing on avoiding water waste and minimizing energy consumption, in the case of the use of pressurized irrigation methods, without losing the quality of the crop. Rational management aims to meet the water needs of crops by supplying water at the right time and in the right quantity, i.e. by adapting the irrigation schedule (PEREIRA, 2004). This procedure can be made efficient and inexpensive with the use of computer software.

Computer software used in irrigation management are programs for microcomputers that calculate crop and irrigation water requirements based on climate, soil and crop data. In addition, computer software allows irrigation schedules to be established for different management conditions, and calculates the water supply scheme of a project for different cropping patterns (PEREIRA, 2004).

One water balance simulation model that has stood out internationally is the ISAREG software, developed at the Instituto Superior de Agronomia in Portugal in 1992, which is capable of dealing with capillary rise and percolation through the root zone. This model has been used in various countries and regions. For the western region of the state of Santa Catarina, ISAREG was validated by Guadagnin et al. (2012). In this study, the model was validated by comparing the results of the reduction in maize yield obtained from the simulation of the water balance using the ISAREG model, with average yield data for this crop in the region of Sao Miguel do Oeste/SC, available from IBGE (2012).

This study was carried out on maize, as it is one of the main agricultural activities in the west of the state of Santa Catarina and is mostly used as feed for pigs. Due to its growing economic importance for agriculture in this region, researchers and research institutions in the region have been looking for corn production systems that are sustainable from the point of view of natural water and soil resources.

Therefore, this work was developed based on the following hypotheses: i) the water needs of the corn crop vary in different soils (Cambissolos, Nitossolos and Latossolos) when subjected to different managements (Cultivo Minimo, Pastagem Perene and Plantio Direto), in the western region of the state of Santa Catarina and can be simulated using the ISAREG model; ii) regional climate modeling can be applied to quantitatively analyze precipitation events in the far west of Santa Catarina with a view to future irrigation planning for corn cultivation in this region.

Given these hypotheses, this work was carried out with the aim of evaluating and quantifying the water requirements of the corn crop in different scenarios in the far west of the state of Santa Catarina using the ISAREG model.

2. literature review

2.1. Water availability in Brazil

Brazil is a country located on the American continent with a territorial extension corresponding to 20.8% of the territory of the Americas and 47.7% of South America. It has an area of approximately 8,500,000 km^2 and is the fifth largest country in terms of land area, surpassed only by Russia, Canada, China and the United States. It is bathed to the east by the Atlantic Ocean and surrounded to the north, west and south by all the countries of the South American continent except Chile and Ecuador (GEO Brasil Recursos Hidricos, 2007).

With regard to climate, 92% of its territory is located in the intertropical zone, a fact which, together with the low altitudes of the terrain, explains the predominance of hot climates, with average temperatures above 20°C. The types of climate found in Brazil are: equatorial, tropical, high altitude tropical, Atlantic tropical, semi-arid and subtropical (GEO Brasil Recursos Hidricos, 2007).

The average annual flow of rivers in Brazil is around 180,000 m^3 /s, which corresponds to approximately 12% of the world's available water resources, which is 1.5 million m^3 /s. If we take into account the flows originating in foreign territory and entering the country (Amazon - 86,321 thousand m^3 /s; Uruguay - 878 m^3 /s and Paraguay 595 m^3 /s), the average flow reaches values of around 267 thousand m^3 /s, corresponding to 18% of the world's water availability. The average flow per inhabitant is around 33,000 m^3 /hab.year, but there is great spatial and temporal variation in flows in the country, where the Amazon Hydrographic Region holds 74% of surface water resources and is inhabited by only 5% of the Brazilian population. The lowest average flow per inhabitant is observed in the Atlantic Northeast East Hydrographic Region, in the northeast of Brazil, with an average of less than 1,200 m^3 /hab-year. In some basins in this region, values of less than 500 m^3 /hab-year are recorded (GEO Brasil Recursos Hidricos, 2007).

2.2. Soil Water Balance in the Soil-Plant-Atmosphere System

Knowing the processes linked to water transfer between plants in the crop's rooting zone, the soil and the atmosphere is fundamental for good management of water resources in irrigation (PEREIRA, 2007). It is therefore necessary to study the water balance in the soil for a given crop. From an agronomic point of view, the water balance is fundamental, as it defines the water conditions under which a crop grows and makes it possible to determine water consumption by plants at different stages of development (REICHARDT and TIMM, 2012).

According to Thornthwaite and Mather (1955), the water balance is defined as an accounting of the inflow and outflow of water in the soil, where the inflow is represented by precipitation and/or irrigation and the outflow by potential evapotranspiration. Van Lier (2010) considers the soil water balance to be the difference between the amount of water entering and the amount of water leaving the soil during a certain period of time. Expressing these quantities in water height, this balance represents the variation in water storage in the volume of soil considered, over the selected time interval. Therefore, knowing the soil's water balance is important for monitoring the amount of water that the soil in question is capable of storing. Reichardt and Timm (2012) describe the water balance as the sum of the amounts of water entering and leaving a soil volume element in a given time interval, resulting in the net amount of water remaining in it.

In a situation where you want to assess the water balance in the soil with an agricultural crop planted in the field, the volume of soil considered will depend on the crop under study, as it must encompass its entire root system. The upper limit of this volume is therefore the soil-atmosphere interface or soil surface and the lower limit is a surface parallel to the former at the depth of the crop's root system. In this case, the quantities of water entering and leaving the soil are represented by the following processes: rainfall (P), irrigation (I), surface runoff (inflow Re and outflow Rs), subsurface runoff (inflow R'e and outflow R's), internal drainage (D), capillary rise (AC) and real evapotranspiration (ET). Equation 1 is then applied to establish the soil water balance:

$$\Delta h = P + I + Re + R'e + Rs + R's + AC + D + ET$$

(Equation 1)

In this equation, Δ h represents the change in water storage in the soil. The values of the input processes are positive and the output processes are negative.

Internal drainage (D) represents the loss of water out of the root zone through the lower limit of the soil volume considered. However, depending on the conditions, instead of leaving, water can enter through this boundary. The entry of water through the lower surface of the soil control volume has been called capillary rise (CA). Surface runoff and subsurface runoff, depending on the position of the area chosen for the balance in the relief of the terrain and the physical conditions of the soil profile, can also be positive or negative increments of water, i.e. entering and leaving laterally over and under the soil surface, respectively (Figure 1).

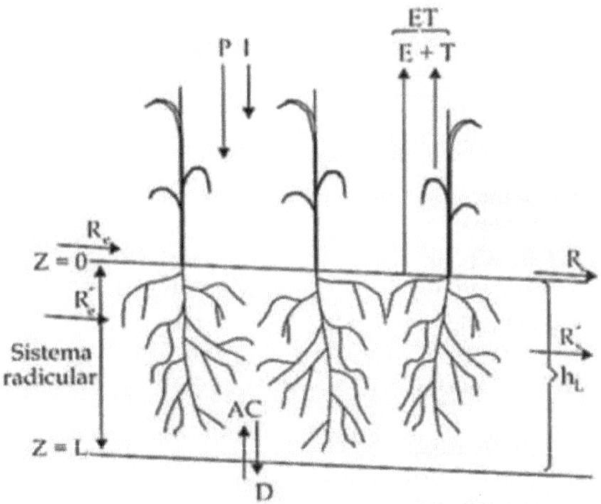

Figure 1 - Soil water balance with an agricultural crop. Source: Van Lier, 2010.

Of course, precipitation (P) and irrigation (I) are input processes and real evapotranspiration (ET) is a water output process, in the volume of soil considered.

In view of this, the quantification of the soil water balance in the root zone of a given crop, with the respective quantification of the terms that make it up and the characterization of the patterns of occurrence of the water transfer processes (patterns

7

of extraction by the roots and of water runoff in the soil) is a determining need for irrigation management and for its improvement and optimization (CAMEIRA et al., 2005).

When determining the variables in the water balance equation, it is relatively easy to determine the terms irrigation (I) and precipitation (P). Precipitation can be quantified using rain gauges and, if irrigation takes place, this liquid liquid can be easily controlled.

2.2.1- Climatological Water Balance

The climatological water balance developed by Thornthwaite and Mather (1955) is used to determine the variation in water storage in the soil without irrigation (I=0). Another simplification, for practical purposes, is to consider capillary rise to be negligible (AC=0). In this way, it is possible to estimate the change in storage, called the change in storage (ALT), the actual evapotranspiration (ETr), and the deep drainage, called the water surplus (EXC), resulting in equation 2:

±ALT = P- ETr- EXC (Equation 2)

In addition to ALT and EXC, the determination of potential or reference evapotranspiration (ETp) and actual evapotranspiration (ETr) allows the estimation of water deficit (DEF), defined as:

DEF = ETp - Etr (Equation 3)

The climatological water balance (CHB) has several applications, including: regional water availability, where regional climatic characterization and comparison of average soil water availability is possible; drought characterization, where calculations are useful in determining periods of drought and their effects on agriculture, such as reduced production. In addition to these, the climatological water balance allows for the determination of the best sowing times, and agro-climatological zoning, which serves as the basis for regional climate studies.

The BHC is most often presented on a monthly scale and for an average (normal) year. However, Thornthwaite and Mather (1955) described that the water balance can also be used to monitor water storage in the soil in real time, i.e. at the moment or even over a given period. This type of water balance is called a sequential or serial water balance and can be done on various time scales such as: daily, weekly, decadal or even monthly. The time scale to be used must be compatible with the purpose of the water balance.

2.2.1.1 - Determining the Soil's Available Water Capacity (AWC)

When preparing the climatological water balance, the first step is to select the CAD, i.e. the water table corresponding to the soil moisture interval between field capacity (CC%) and the permanent wilting point (PMP%). The water balance, according to Thornthwaite and Mather (1955), is most commonly used to characterize the water availability of a region on a climatological and comparative basis. The CAD is selected according to the type of crop to which it is to be applied, rather than the type of soil.

The CAD (mm) must be determined according to the soil's physical and hydraulic properties, such as: CC%, PMP%, relative soil density (Dr) and effective depth (Z (mm)) of the root system of the plants under cultivation:

$$\mathbf{CAD = 0,01 * (CC\% - PMP\%) * Dr * Z}$$

(Equation 4)

Effective depth is the depth at which around 80% of the roots are concentrated, expressed in mm. This depth depends not only on the type of soil, but also on the crop and the hydric and nutritional regime to which the plant is subjected.

The physico-hydric properties of soil depend on its texture and structure and are highly variable.

Ideally, the DAC should be calculated for the local soil and crop conditions, including taking into account how the depth of the root system varies with the growth stage of the crop. If there is a marked variation in physical-hydric properties with depth, the DAC of each soil layer should be calculated, with the total DAC of the effective depth given by the sum of the DAC of each layer.

2.2.2- Crop water balance

The water balance described above, called climatological, aims to understand the conditions of the water balance in soil covered by standard vegetation, since ETp or ETr are, by definition, characteristic of an extensive area of grass, actively growing, completely covering the soil, with a height of between 8 and 15 cm, respectively, without and with water restrictions.

In the case of a crop-specific water balance, the aim is to calculate water storage in the soil, taking into account both the type of vegetation and its stage of growth and development. In this situation, the plant does not always completely cover the ground and its leaf area (transpiring surface) varies with age (days after planting or emergence). Under these conditions, when there is no water deficit, evapotranspiration differs from potential evapotranspiration and is called maximum crop evapotranspiration, or simply crop evapotranspiration (ETc), which will be taken into account in the crop's water balance. Due to the difficulty of measuring crop evapotranspiration, it is more convenient to calculate it as a function of ETo, according to Jensen (1968):

$$\mathbf{ETc = Kc * ETo} \qquad \text{(Equation 5)}$$

Therefore, the estimate of ETc depends on an adjustment coefficient (Kc), called the crop coefficient. This, in turn, is a function of the crop's leaf area index (LAI), which varies with its growth and development, i.e., it is directly related to the size of its transpiring leaf surface, since the larger the leaf area, the larger the transpiring surface, and the greater the plant's potential for using water.

Table 1 shows the average Kc values for different crops in the different phenological phases. For annual crops, there is a gradual increase in Kc values up to the flowering stage, which coincides with the time of maximum leaf area, decreasing from the end of fruiting and the beginning of ripening, as a result of leaf senescence.

For most crops, Kc ranges from 0.3 in the crop establishment phase to 1.2 in the flowering and fruiting phase, with maximum Kc values hardly exceeding 1.2.

Table 1- Crop coefficient (Kc) for some crops. Source: Doorenbos and Kassan (1994).

Culture	Stages of Culture Development				
	Establish.	Des. Veget.	Floresc.	Fruiting	Maturation
Alfalfa	0,3 - 0,4	-	-	-	1,05 -1,2
Cotton	0,4 - 0,5	0,7 - 0,8	1,05 - 1,25	0,8 - 0,9	0,65 - 0,7
Peanuts	0,4 - 0,5	0,7 - 0,8	0,95 - 1,1	0,75 - 0,85	0,55 - 0,6
Rice	0,4 - 0,5	0,7 - 0,8	0,9 - 1,2	0,8 - 0,9	0,5 - 0,6
Tropical Banana	0,4 - 0,5	0,7 - 0,85	1,0 - 1,1	0,9 - 1,0	0,75 - 0,85
Subtropical banana	0,5 - 0,65	0,8 - 0,9	1,0 - 1,2	1,0 - 1,15	1,0 -1,15
Potato	0,4 - 0,5	0,7 - 0,8	1,05 - 1,2	0,85 - 0,95	0,7 - 0,75
Sugar beet	0,4 - 0,5	0,75 - 0,85	1,05 - 1,2	0,9 - 1,0	0,6 - 0,7
Sugar cane	0,4 - 0,5	0,7 - 1,0	1,0 - 1,3	0,75 - 0,8	0,5 - 0,6
Dried onion	0,4 - 0,6	0,7 - 0,8	0,95 - 1,1	0,85 - 0,9	0,75 - 0,85
Green onions		0,6 - 0,75	0,95 - 1,05	0,95 - 1,05	0,95 - 1,05
Café com Trato			0,65 - 0,8		
Untreated coffee			0,85 - 0,9		
Treated citrus			0,65 - 0,75		
Untreated citrus			0,85 - 0,9		
Peas	0,4 - 0,5	0,7 - 0,85	1,05 - 1,2	1,0 - 1,15	0,95 - 1,1
Green Beans	0,3 - 0,4	0,65 - 0,75	0,95 - 1,05	0,9 - 0,95	0,85 -0,95
Dried Beans	0,3 - 0,4	0,7 - 0,8	1,05 - 1,2	0,65 - 0,75	0,25 - 0,3
Sunflower	0,3 - 0,4	0,7 - 0,8	1,05 - 1,2	0,7 - 0,8	0,35 - 0,45
Watermelon	0,4 - 0,5	0,7 - 0,8	0,95 - 1,05	0,8 - 0,9	0,65 - 0,75
Sweet Corn	0,3 - 0,5	0,7 - 0,9	1,05 - 1,2	1,0 - 1,15	0,9 - 1,1
Grain corn	0,3 - 0,5	0,7 - 0,85	1,05 - 1,2	0,8 - 0,95	0,55 - 0,6
Oliveira			0,4 - 0,6		
Green Pepper	0,3 - 0,4	0,6 - 0,75	0,95 - 1,1	0,85 - 1,0	0,8 - 0,9

2.2.3- Water balance for irrigation control

Irrigation is an agricultural operation to meet the water needs of crops, and is fundamental in production systems in regions with regular droughts (Pereira, 2002). It provides an important degree of stability for food production, since weather adversities are minimized. In the case of excessive irrigation, there are losses related to unnecessary fuel or electricity costs, degradation of soil quality, leaching of essential plant nutrients and even reduced productivity. In view of this, the possibility of correctly estimating evapotranspiration and from these estimates determining the amount of water to be supplied to the soil is of considerable importance for monitoring irrigation.

The water balance for irrigation control is an adaptation of the sequential climatological water balance in order to facilitate its application in field conditions, without the need for sophisticated computational resources, by measuring only rainfall and the meteorological elements required in the method chosen to estimate reference evapotranspiration (ETo).

Before starting an irrigation project, you need to know fundamental aspects such as the phenology of the crop, which is directly linked to water requirements and its phenological phases, basically related to the crop's Kc, the crop's water demand, related to weather conditions, mainly available liquid radiation and atmospheric demand and ETc, which can be conveniently estimated as a function of ETo and finally, the physical characteristics of the soil profile, which are necessary to determine the volume of water available to the roots.

2.2.3.1 Irrigation monitoring roadmap

In order to monitor irrigation with a climatological water balance, it is necessary to pre-determine the irrigation rate (DR, mm) or Irrigation Limit (LI, mm), which corresponds to the amount of water to be applied for each irrigation. The irrigation rate (DR) can be fixed, i.e. the same value is always used for each irrigation, or variable. In the first case, the irrigation rate is pre-set (fixed DR), varying between a minimum value of readily available water (0.25AFD) and a maximum (0.50AFD). In the second case, the irrigation rate is variable (variable DR), always seeking to raise the water storage in the soil to field capacity. Therefore, what differentiates the two criteria is the way in which the volume of water to be applied at the time of irrigation is calculated. Figures 2 and 3 illustrate irrigation monitoring for a hypothetical crop, taking into account the two criteria mentioned above (PEREIRA et al., 2002).

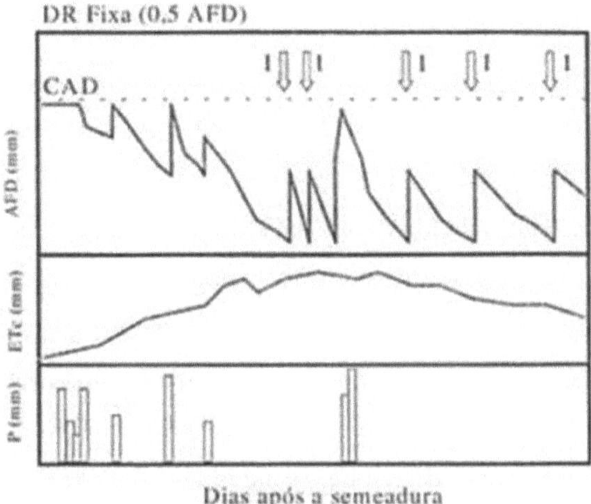

Figura 2 - Schematic representation of irrigation monitoring for a hypothetical crop, considering the adoption of fixed irrigation (fixed DR). Source: Pereira (2002).

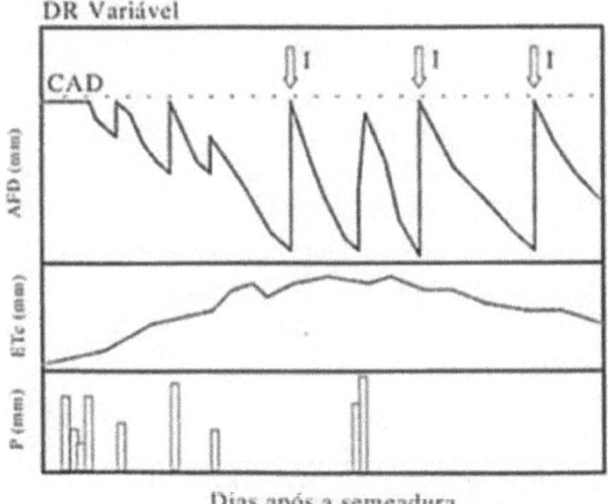

Figura 3 - Schematic representation of the irrigation monitoring of a hypothetical crop, considering a variable irrigation allowance (variable DR). Source: Pereira (2002).

It can be seen that, for the same conditions of rainfall (P) and crop evapotranspiration (ETc), the frequency of irrigation is higher in the fixed DR, but with lower rates than in the variable DR, which has a lower number of irrigations, but with higher rates in order to re-establish the soil's water storage at field capacity.

Therefore, the steps for monitoring irrigation using the climatic water balance are:

13

1) **Determine the CAD using the following equation:**

$$CAD = 0,01 * (CC\% - PMP\%) * D * Z$$

(Equation 6)

2) **Determining easily available water (AFD)**

When you have an irrigation system, you shouldn't wait until the plants show the external symptoms of a lack of water before you irrigate. If this happens, production will already be affected. Therefore, irrigation should be started before the plants reach this point. In practice, this point represents a fraction (percentage p) of the CAD, called Easily Available Water (AFD), i.e. the water that can be extracted from the soil from maximum storage, without a water deficit occurring in the crop. AFD is calculated by:

$$AFD = p * CAD$$

(Equation 7)

The p fraction is determined experimentally and is a function of the type of crop and the maximum water consumption at the different phenological stages (table 2). This results in different AFD values during the crop cycle, which makes it difficult to calculate the water balance. For practical purposes, p= 0.35 is usually adopted for crops in groups 1 and 2, and p= 0.50 for crops in groups 3 and 4, listed in Table 2.

Table 2 - p-fraction for ETc crop groups. Source: Doorenbos and Kassan (1994)

Cultures	Group	ETc (mm d-)^1								
		2	3	4	5	6	7	8	9	10
Onion, Pepper, Potato	1	0,50	0,43	0,35	0,30	0,25	0,23	0,20	0,20	0,18
Potatoes, cabbage, grapes, Peas, Tomatoes	2	0,68	0,58	0,48	0,40	0,35	0,33	0,28	0,25	0,23
Alfalfa, Beans, Citrus, Peanut, Sunflower, Wheat	3	0,80	0,70	0,60	0,50	0,45	0,43	0,38	0,35	0,30
Cotton, Maize, Sorghum, Soy,										

---

3) Determining crop evapotranspiration ETc

Equation 8 is used to determine ETc:

$$ETc = Kc * ETo$$

(Equation 8)

4) Precipitation (P)

This is the total observed rainfall (mm) over the period in question. It is important to measure it in the area to be irrigated, as it is a meteorological element with a lot of spatial variability and discontinuity.

5) Irrigation (I)

Irrigation means the amount of water to be applied at the beginning of the period in question, and whenever the AFD at the end of the previous period (AFDf) is close to the critical limit, i.e. AFD$\approx$ 0. The amount of irrigation water depends on the criterion adopted (fixed or variable DR).

6) Initial easily available water (AFDi)

This is the AFD at the start of the period in question.

- When there is no irrigation: AFDi of the period = AFDf of the previous period.

- When there is fixed DR irrigation: AFDi of the period = I + AFDf of the previous period.

- When there is irrigation with variable DR: AFDi of the period = AFDf of the previous period.

7) Final readily available water (AFDf)

This is the AFD at the end of the period, resulting from the following balance sheet:

- For fixed DR:

$$AFDf = AFDi + (P - ETc)$$

- For variable DR:

$$AFDf = AFDi + (I + P - ETc)$$

2.3- Evapotranspiration (ET)

Water consumption by the crop is technically called evapotranspiration (RADIN et al., 2000). According to Mendonça et al. (2003), evapotranspiration is the inverse of precipitation, as it is the sum of water loss through soil evaporation and plant transpiration, and is controlled by the energy balance, atmospheric demand and the supply of water from the soil to the plants.

According to FAO Bulletin 56, evapotranspiration can be conceptualized as the combination of two processes, evaporation and transpiration, by which water is lost to the atmosphere. The evaporation of water from the soil is the process by which water is lost from the soil to the atmosphere, it is the process by which water, after vaporization, is removed from the evaporation surface. It evaporates from various surfaces, such as lakes, rivers, soils and even wet vegetation. Transpiration, on the other hand, is the process of water loss from the plant to the atmosphere through transpiration, i.e. transpiration is a biophysical process by which water, which has been part of the plant's metabolism, is transferred by vaporization to the atmosphere through the stomata.

According to Bulletin 56 (FAO), transpiration and evaporation vary according to the energy supply. Therefore, incident solar radiation, temperature, relative humidity, wind intensity and speed, the size of the evaporating or transpiring surface, soil texture and the depth of the water table are all factors that contribute to determining the degree of evapotranspiration that occurs in the soil. Other factors that influence the rate of evapotranspiration are related to environmental aspects, cultivation practices and also crop characteristics such as: phenological development stage and leaf area index.

As it is practically impossible to distinguish between *water* vapor from evaporation of water in the soil and transpiration from plants, evapotranspiration is therefore defined as the simultaneous process of water transfer to the atmosphere by evaporation of water from the soil and wet vegetation and by transpiration from plants.

Measuring the water lost through evapotranspiration is of great importance in determining the water needs of crops, as this is the amount of water that must be replaced in the soil by rain or irrigation.

2.3.1- Evapotranspiration: definitions

Before describing some methods capable of estimating the evapotranspiration that occurs in the environment, it is extremely important to know and characterize the definitions of evapotranspiration.

Reference evapotranspiration (ETo) - according to Doorenbos and Pruitt (1977), ETo is defined as the evapotranspiration that takes place in a soil covered entirely by grass (reference surface), with active and uniform growth, without water deficiency and with a height of between 8 and 15 cm.

According to Bulletin 56 (FAO), the reference surface corresponds to a hypothetical grass crop with specific characteristics. The concept of reference evapotranspiration is introduced to study the evapotranspiration demand of the atmosphere, independent of the type and development of the crop and local management practices. As it refers to evapotranspiration under the same reference surface, it is possible to compare ETo values measured or estimated in different locations and at different times of the year.

The only factors that affect ETo are climatic parameters. Therefore, ETo can be calculated from them. Therefore, this is an indicative value of the evapotranspirative demand of the atmosphere in a given location over a given period, without taking into account plant characteristics or soil factors.

Crop evapotranspiration (ETc) - is the amount of water used by a crop at any stage of its development, from planting (sowing) to harvest, when there are no water restrictions.

ETc varies according to leaf area (transpiring surface), as the larger the leaf area, the higher the ETc for the same atmospheric demand.

Actual evapotranspiration (ETr) - can be defined as the amount of water actually used by an extensive surface vegetated with grass, actively growing, completely covering the ground, but with or without water restrictions.

When there are no water restrictions ETr= ETo, but when there are water restrictions ETr$\leq$ ETo.

2.3.2- Methods for estimating evapotranspiration (ET)

Various methods for estimating evapotranspiration are available, basically divided into two groups: direct methods and indirect or empirical methods.

Direct methods are characterized by the determination of evapotranspiration directly in the area, where the different types of Iisimeters stand out, or even by the soil *water* balance method. Indirect methods are characterized by the use of empirical equations or mathematical models, which use weather and climate-physiological data for their application. As these are estimates, they are less accurate, especially when applied to climatic conditions other than those under which they were developed.

The choice of method for estimating evapotranspiration for a given location will depend on the availability of climatic elements. In practical terms, before choosing the method to be used, it is necessary to know which climatic elements are available for the study. From there, you determine which methods can be applied.

2.3.2.1   - Direct measurement of evapotranspiration using isometers

As the name implies, this method aims to measure evapotranspiration using lysimeters, either directly or using the soil *water* balance method.

Lysimeters are buried tanks containing a representative sample of the soil and vegetation you want to study, and they must represent real field conditions fairly faithfully. The plants inside the lysimeter must be similar to those around it in all agronomic aspects: variety, stage of development, physical and health conditions, fertilization, etc.

There are different types of lysimeters: drainage lysimeters and mechanical or electronic weighing lysimeters. Figure 4 shows a schematic of a drainage or percolation lysimeter.

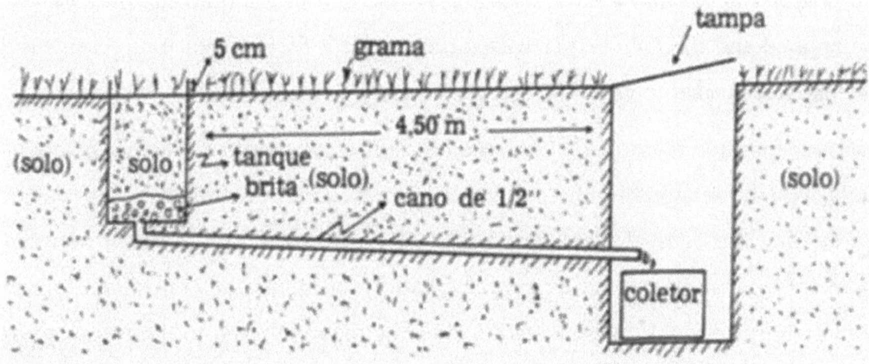

Figure 4 - Schematic of a drainage or percolation isimeter. Source: Pereira, (2002).

Lysimeters are devices that allow evapotranspiration to be determined by the difference, for a given period of time, between the water supplied and the water percolated.

The most commonly used are:

Weighing lysimeters: use the automated measurement of load cells installed under a waterproof box, measuring the variation in its weight. In this way, when water is consumed by the plants in the lysimeter, the weight of the control volume decreases, which is proportional to evapotranspiration (GOMIDE et. al., 1996).

Constant water **table lysimeter**: this adopts an automatic system for feeding and recording the replenished water, in order to keep the water table level constant, with evapotranspiration being equal to the volume of water leaving the feeding system (CAMARGO, 1962).

Drainage lysimeters - work well over long periods of observation and are based on the principle of conserving the mass of water in a volume of soil (CAMARGO,

1962).

According to Bueno (2012), when operating a drainage or percolation lysimeter (Figure 4), water is added in sufficient volume to allow drainage. When drainage ceases, it can be guaranteed that the moisture content of the soil inside the tank is at its maximum retention capacity. After a certain period of time, water is added to the tank, also in a volume that allows drainage. At the end of the drainage process, the volume applied and the volume percolated are counted; the difference represents the volume that was needed to bring the soil back to its maximum retention capacity.

As evapotranspiration is normally expressed in millimeters per day, it is sufficient to divide the volume retained by the surface area of the tank, obtaining the equivalent blade. Dividing the equivalent blade by the volume retained and the time between measurements gives the average evapotranspiration for the period considered for the crop under study. The following equation uses the data collected in the lysimeter to represent the evapotranspiration of the crop in question.

$$ETc = \frac{\dfrac{(Va - Vp)}{A} + P}{T}$$

(Equation 9)

Where:

ETc = average crop evapotranspiration [mm d];[-1]

Va = volume of water applied [L];

Vp = volume of percolated water [L];

A = tank area [m];[2]

T = interval between measurements [day]; and

P = rainfall during the period considered [mm].

2.3.2.2 - Empirical or indirect methods for estimating reference evapotranspiration (ETo)

Estimating evapotranspiration using mathematical equations is the most common and usual way of obtaining the water loss observed in a vegetated area. ETo values can be estimated from elements measured at weather stations, and there are several methods for this.

Generally speaking, all the methods are empirical, because for them to be fully applied, some empirical parameterization is required. The main methods for estimating reference evapotranspiration are described below:

2.3.2.2.1- Class A evaporimetric tank method

According to Pereira (2002), the evaporimetric tank or "Class A" tank method was developed to provide a practical way of estimating ETo, applied to irrigation management. The tank used in this process is small (Figure 5) with the sides directly exposed to solar radiation, and the water in the tank does not hinder the evaporative process and is always available, even during dry periods. Therefore, the evaporation value obtained in the tank is exaggerated in relation to the actual loss of a crop, even if it is under optimum soil water supply conditions. Therefore, the daily evapotranspiration value measured in the tank (ECA mm day^{-1}) needs to be corrected by an adjustment factor, called the tank coefficient (kp), in order to obtain the corresponding ETo. This can be obtained using equation 10 below:

$$\textbf{ETo} = \textbf{Kp} * \textbf{ECA} \qquad \text{(Equation 10)}$$

Where:

ETo= evapotranspiration of the reference crop [mm d⁻ 1];

Kp = tank coefficient;

ECA = evaporation measured in the "Class A" tank [mm d⁻ i];

Figure 5- "Class A" evaporimetric tank. Source: Pereira (2002).

In order to determine the evaporation occurring in the tank (ECA), a hook micrometer is used and the readings are taken in a tranquilizer well installed inside the tank, the purpose of which is to prevent disturbances on the liquid surface, especially small waves formed by winds, and thus enable the water level to remain stable when the readings are taken.

The value of Kp, which is always less than 1, depends on the conditions of relative humidity (RH, in %), wind speed (V, in km d^{-1}) and the length of the border (L, in m), vegetated or not, surrounding the tank. This method is recommended by the FAO (Food and Agricultural Organization) (DOORENBOS AND KASSAM, 1994), and Kp values can be found in Table 3.

Table 3 - Class A Tank coefficient (Kp), for different borders of low vegetation around the tank, and levels of relative humidity and wind speed over 24 hours. Source: Doorenbos and Kassan (1994).

Wind (Km/day)	Edge (m) -	Relative humidity		
		< 40%	40% a 70%	> 70%
	1	0.55	0.65	0.75
Lightweight	10	0.65	0.75	0.85
< 175	100	0.70	0.80	0.85
	1000	0.75	0.85	0.85
	1	0.50	0.60	0.65
Moderate	10	0.60	0.70	0.75
175 a 425	100	0.65	0.75	0.8
	1000	0.70	0.80	0.8
	1	0.45	0.50	0.6

Strong	10	0.55	0.60	0.65
425 a 700	100	0.60	0.65	0.7
	1000	0.65	0.70	0.75
	1	0.40	0.45	0.50
Very Strong	10	0.45	0.55	0.60
> 700	100	0.50	0.60	0.65
	1000	0.55	0.60	0.65

2.3.2.2.2- Estimating reference evapotranspiration using the Hargreaves method

This is because the Penman-Monteith FAO-56 method requires accurate measurements of air temperature, relative humidity, solar radiation and wind speed. Unfortunately, there are only a limited number of weather stations where these variables are measured efficiently, especially in central Brazil. Therefore, the search for alternative methods that require a smaller number of meteorological variables to estimate ETo has been a viable solution to overcome this problem. One alternative method that has been used in several studies is the Hargreaves equation (HARGREAVES; SAMANI, 1985). Although this equation is used to estimate ETo for weekly or longer periods, there are studies that illustrate that this adjusted equation can accurately estimate daily ETo. This equation estimates ETo based solely on air temperature data. However, it requires local calibration for acceptable performance. This calibration must be carried out by adjusting the coefficients of the Hargreaves equation. This is because calibrations using only linear regression models overestimated ETo values when compared to the values obtained by Penman-Monteith (ALLEN et al., 1998).

According to Pereira et al. (2002), the Hargreaves and Samani method for estimating ETo was developed in 1985 for the semi-arid conditions of the State of California, USA (city of Davis), based on the evapotranspiration obtained in a weighing lysimeter cultivated with grass. The advantage of this method is that it can be applied in arid and semi-arid climates, such as in the north-east of Brazil, but it has limitations for use in regions with a humid climate, as it overestimates the results obtained. This method is represented by equation 11 below:

$$ETo = 0,0023 * (Tar + 17,8) * (Tmax - Tmin)^{0,5} * Ra$$

(Equation 11)

Where:

ETo= reference evapotranspiration [mm.day];[-1]

Tar = daily average air temperature [°C];

Tmax= daily maximum air temperature [°C];

Tmin= daily minimum air temperature [°C];

Ra= daily total top-of-atmosphere solar radiation [mm.day⁻ i].

2.3.2.2.3- Estimation of reference evapotranspiration through of the Thornthwaite method

This was one of the first methods developed exclusively to estimate reference evapotranspiration (THORNTHWAITE, 1948).

The Thornthwaite method estimates the monthly reference evapotranspiration (ETo), taking as a standard a 30-day month with twelve hours of daily sunshine.

The Thornthwaite method requires the average temperature for each month as input data, according to equation 12:

$$ETo = 16 \frac{(10*Ti)^{a}}{I} \quad (0 < Ti < 26,5\ °C)$$

(Equation 12)

Where:

ETo= reference evapotranspiration [mm.day-];[1]

Ti= average monthly temperature for month i (°C);

I= heat index of the region;

a= coefficient, also related to temperature.

The heat index can be calculated using equation 13:

$$I = \sum_{i=1}^{12} \left(\frac{Ti}{5}\right)^{1,514}$$

(Equation 13)

The coefficient **a** can be calculated using equation 14 below:

$$a = 6,75.10^{-7}.I^3 - 7,71.10^{-5}.I^2 + 1,7912.10^{-2}.I + 0,49239$$

(Equation 14)

These Iea coefficients, calculated using climatological normals, are characteristic of the region and become constant, being independent of the year in which ETo is estimated.

The ETo value calculated, by definition, represents the total monthly evapotranspiration that would occur under those thermal conditions, but for a standard 30-day month, in which each day would have 12 hours of photoperiod. Therefore, in order to obtain the ETo for the corresponding month, this ETo value must be corrected according to the actual number of photoperiod days in the month.

2.3.2.2.4- Estimating reference evapotranspiration using the Penman-Monteith method

This is a micrometeorological method, described by Monteith (1965), which was adapted by Allen et al. (1989) to estimate reference evapotranspiration (ETo) on a daily scale.

The Penman-Monteith method, currently adopted as the standard method by the FAO (PM-FAO), was an improvement on the original Penman method (PEN-1948), and is one of the various existing methods for calculating ETo and is considered by several researchers to be the most accurate, as the most accurate method because, as well as trying to consistently represent the biophysical phenomenon of evapotranspiration, it is fed by almost all the meteorological elements observed at surface weather stations (SMITH et al., 1991).

Using the Penman-Monteith FAO method, it is possible to calculate both the hourly evapotranspiration and then the daily evapotranspiration by adding up the hourly values, and to calculate the daily evapotranspiration using the average daily data.

This method can be represented by equation 15 below:

$$ET_o = \frac{0,408.\Delta.(R_1 - S) + \gamma \frac{900}{T_{ar} + 273}.U_2.(e_s - e_a)}{\Delta + \gamma(1 + 0,34.U_2)}$$

(Equation 15)

Where:

ETo= reference evapotranspiration [mm.day];$^{-1}$

Δ= tangent to the vapor pressure curve [kPa.°C⁻ i];

R₁= Radiation balance or net radiation [MJ.m^{-2} .day⁻ i];

S= sensible heat flux to the ground [MJ.m⁻ 2.dia⁻ i];

Y= psychrometric constant [kPa.°C⁻ i];

T_{ar} = average daily air temperature measured 2 meters above the ground surface [°C];

U_2 = wind intensity, daily average at 2 meters from the ground surface [m.s⁻ i];

es= saturation vapor pressure [kPa];

ea= current vapor pressure [kPa].

2.4-- Growing corn

Maize (*Zea mays L.*), which belongs to the Poaceae family, is one of the most important cereals grown and consumed in the world due to its production potential, chemical composition and nutritional value. Due to its many applications, both in human food and animal feed, it plays an important socio-economic role, as well as being an indispensable raw material for diversified agro-industrial complexes (FANCELLI and DOURADO NETO, 2000).

Maize is the most important commercial plant to have originated in the Americas.

There are indications that it originated in Mexico, Central America or the southwest of the United States. It is one of the oldest crops in the world, and there is evidence from archaeological and geological excavations and radioactive disintegration measurements that it has been cultivated for at least 5,000 years. Soon after the discovery of America, it was taken to Europe, where it was grown in gardens until its food value became known. It was then planted on a commercial scale and spread from latitude 58° north (Soviet Union) to 40° south (Argentina) (GODOY, 2002).

The use of corn grain as animal feed accounts for most of the consumption of this cereal, i.e. around 70% worldwide. In the United States, around 50% is used for this purpose, while in Brazil it ranges from 60 to 80%, depending on the source of the estimate and from year to year (DUARTE, 2000).

Although it doesn't play a very large role in the use of corn grain, human nutrition with corn derivatives is also an important factor in the use of this cereal, especially in low-income regions. In some situations, maize is the daily staple food, such as in the northeast of Brazil, where maize is the source of energy for many people living in the semi-arid region; another example is the Mexican population, which uses maize as a basic ingredient in its cuisine.

In Brazil, corn is one of the most traditional crops, occupying significant positions in terms of its value in agricultural production, cultivated area and volume produced, especially in the South, Southeast and Midwest regions of the country (FNP, 2002).

Brazil is the world's third largest corn producer, surpassed only by the United States and China. According to CONAB data from May 2012, the corn harvest and the national cultivation of the cereal reached record levels both in terms of area and production this year. In 2012, the corn crop reached an index of 65,903.7 tons harvested, which represents a percentage increase of 14.8% when compared to the previous year's crop. Of the total crop harvested in 2012, the southern states of the country contributed 31.82%, with the state of Santa Catarina accounting for 4.55% of the total percentage produced in the 2011/2012 harvest. This is due to various factors such as: an increase in the area sown, the use of technology, favorable weather conditions in most of the producing states, precision farming and the use of high-tech seeds.

2.4.1 - Corn crop phenology

Maize has a varied vegetative cycle, ranging from extremely precocious genotypes, whose pollination can occur 30 days after emergence, to those whose life cycle can reach 300 days. However, in our climatic conditions, the maize crop has a variable cycle of between 110 and 180 days, depending on the characterization of the genotypes (super-early, early and late), which is the period between sowing and harvesting (FANCELLI and DOURADO NETO, 2000).

Generally speaking, the crop cycle comprises the following stages of development: (i) germination and emergence: the period from sowing to the actual appearance of the seedling, which, depending on the temperature and humidity of the soil, can last from 5 to 12 days; (ii) vegetative growth: the period from the second leaf to the start of flowering, which varies in length. This is commonly used to characterize maize genotypes in terms of cycle length; (iii) flowering: the period between the start of pollination and the start of fruiting, which rarely lasts more than 10 days; (iv) fruiting: The period from fertilization to the complete filling of the grains, estimated to last between 40 and 60 days; and (v) maturity: the period between the end of fruiting and the appearance of the black layer, relatively short and indicative of the end of the plant's life cycle.

However, for ease of management and study, as well as with the aim of establishing correlations between physiological, climatological, phytogenetic, entomological and phytopathological elements with plant performance, the maize crop cycle was divided into 11 distinct stages of development, according to Fancelli (1986), cited by Fancelli and Dourado Neto (2000) (Figure 6) (i) stage 0 (from sowing to emergence); (ii) stage 1 (plant with four fully unfolded leaves); (iii) stage 2 (plant with eight leaves); (iv) stage 3 (plant with twelve leaves); (v) stage 4 (shoot emission); (vi) stage 5 (flowering and pollination); (vii) stage 6 (milky grains); (viii) stage 7 (pasty grains); (ix) stage 8 (beginning of the formation of the "teeth" which is the concavity in the upper part of the grain); (x) stage 9 ("hard" grains); and (xi) stage 10 (physiologically mature grains).

However, it should be noted that the stages of growth and development prior to the appearance of the ears are identified by assessing the number of fully expanded or unfolded leaves. Thus, a maize leaf can be considered to be unfolded when it has an easily visible lamina-basin junction line ("collar"). For the stages after the ear has emerged, identification should be based on the development and consistency of the kernels (KINIRY and BONHOMME, 1991).

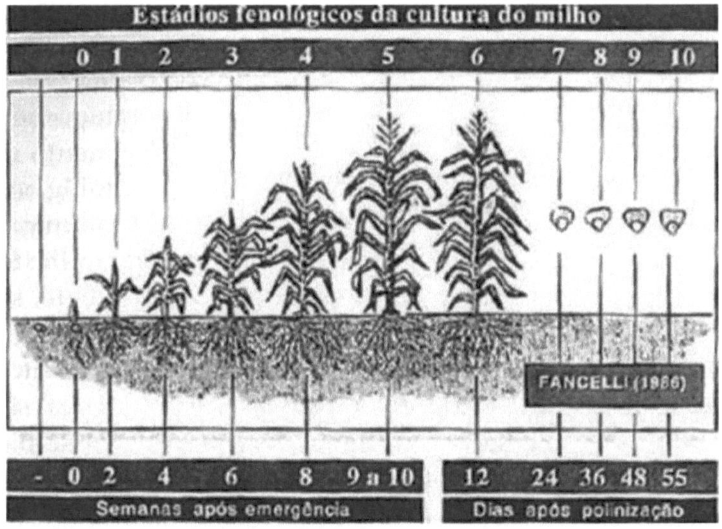

Figure 6 - Crop cycle: corn development stages. Source: Fancelli, 1986, adapted from Nel and Smit, (1978).

2.4.2 - Factors that can affect corn yields

Compared to other cultivated species, maize has experienced significant advances in the most diverse areas of agronomic knowledge, as well as in ecology and ethnobiology, providing a better understanding of its relationships with the environment and man. These interactions are fundamental for predicting plant behavior when subjected to stimuli and negative actions arising from the action of biotic and abiotic agents in the production system (FANCELLI and DOURADO NETO, 2000). In view of this, it is important to know how to expand our knowledge of the maize plant and its relationship with certain climatic conditions and factors that can contribute to or hinder the productive development of this crop.

As a plant of tropical origin, maize requires heat and humidity during its growing cycle in order to develop and produce satisfactorily, providing rewarding yields.

Regardless of the technology applied, the period of time and the climatic conditions to which the crop is subjected are preponderant production factors. Temperature and rainfall are among the most widely studied climatic elements used to assess the feasibility and timing of various agricultural activities.

The processes of photosynthesis, respiration, transpiration and evaporation are direct functions of the energy available in the environment, commonly referred to as heat; while the growth, development and translocation of photoassimilates are linked to the availability of water in the soil, and their effects are more pronounced in conditions of high temperatures where the rate of evapotranspiration is high (FANCELLI and DOURADO NETTO, 2000).

Experimental evidence shows that temperature is one of the most important and decisive factors in the development of maize, although water and other climatic components have a direct influence on the process (ANDRADE, 1992).

At the time of sowing, it is important to note that the soil should be at a temperature above 10°C, with humidity close to field capacity (FC), enabling the emergence process to take place. During plant growth and development, the air temperature should be around 25°C and be associated with adequate water availability in the soil and plenty of light. In the flowering and grain filling period, favorable temperature and light, high availability of water in the soil and relative humidity of more than 70% are the basic requirements for good plant development, but for the harvest period, a predominantly dry climate is necessary (FANCELLI and DOURADO NETTO, 2000).

Regions whose summers have average daily temperatures below 19°C and nights with average temperatures below 12.8°C are not recommended for growing maize.

Soil temperatures below 10°C and above 42°C significantly impair germination, while those between 25 and 30°C provide the best conditions for seed germination and seedling emergence. During flowering and ripening, average daily temperatures above 26°C can accelerate these phases, while temperatures below 15.5°C can rapidly slow them down (BERGER, 1962).

With regard to light, the maize crop responds with high yields to increasing light intensities, as it belongs to the "C4" group of plants, which gives it high biological productivity. Maize is originally a short-day plant, although the limits of these hours of light are neither identical nor well defined for the different cultivars. The occurrence of long days can lead to an increase in the vegetative phase and the number of leaves, causing flowering to be delayed (FERRAZ, 1966). 30 to 40% of light intensity in maize causes a delay in the maturation of organs, especially in late cultivars, which are more sensitive to light deficiency. The greatest sensitivity to light variation is seen at the beginning of the reproductive phase, i.e. in the period corresponding to the first 10-15 days after flowering. At this stage, the reduction in the availability of light radiation causes a reduction in grain density (specific mass) (FANCELLI and DOURADO NETO, 2000).

With regard to wind, its incidence on corn crops can increase the plant's demand for water, making it more susceptible to short periods of drought, as well as promoting lodging of the crop. Similarly, cold or hot winds can cause pollination failures, often becoming an important limiting factor in corn production in some regions (ARNON, 1975).

Initial water deficiencies can significantly affect the germination process and compromise the establishment of the crop, while later deficiencies can paralyze growth and delay the reproductive development of the plants (FANCELLI and DOURADO NETO, 2000). The greatest water requirements are concentrated in the emergence, flowering and grain formation phases. However, in the period between 15 days before and 15 days after the appearance of the male inflorescence, the requirement for a satisfactory water supply combined with suitable temperatures makes this period extremely critical. This is why this phase should be carefully planned to coincide with a season with favorable temperatures (25 to 30°C) and frequent rainfall (FRATTINI, 1975).

The maize crop requires a minimum of 350-500 mm of rainfall in the summer for good productivity, without the need for irrigation. The occurrence of slight water stress at the beginning of the crop's development can stimulate greater plant root development, as long as the soil below 20 cm from the surface is supplied with available water. Periods of water deficiency of one week at the time of setting can cause a drop in production of around 50%, while under the same conditions, water deficiency after pollination will cause damage of between 25 and 32%.

Water consumption by maize in a hot, dry climate rarely exceeds 3.0 mm/day while the plant is less than 30 cm tall. However, during the period between spike and maturity, consumption can rise to 5.0 to 7.5 mm/day (DAKER, 1970).

The amount of water available to the crop depends on the depth explored by the roots, the soil's water storage capacity and the plant's root density. Therefore, rational soil and crop management is of the utmost importance for the growth and distribution of the root system, favoring the efficient use of water in the production process.

There are marked differences in the effective depth of soil solution extraction by the roots of annual crops. In tropical regions, the effective depth of roots rarely exceeds 30 cm and, in temperate climates, it can reach values of over 1.0 m.

However, in order to estimate the humidity factor in a region's climate, rainfall information is not enough. It is essential to consider water losses through evapotranspiration (FANCELLI and DOURADO NETO, 2000).

2.5- Water use in the state of Santa Catarina

Over time, watercourses in Santa Catarina have undergone a substantial change in use, from simple means of transportation and artisanal fishing to diversified uses, in line with the economic development observed in recent decades. The following are some of the main uses of the state's water resources.

In the state of Santa Catarina, the main uses of water resources involving water diversion are associated with human supply, industrial supply, animal watering and irrigation.

With regard to human water supply, data from the Santa Catarina State Water and Sanitation Company (CASAN) shows that all 293 municipalities in the state are supplied with water for human consumption. According to CASAN estimates, in 2002, the urban population of the state served by the water supply system was 4,133,124 inhabitants, which generated a human consumption of drinking water in the state of Santa Catarina of around 25 million m^3 /month. The demand for human supply in the rural areas of Santa Catarina is around 3.2 million m3/month, and is slightly more pronounced in the Atlantic Strand (51.4%) than in the Inland Strand (48.53%). Hydrographic Region 2 - Midwest, located on the Inland Slope, has the second highest rural demand in the state (14.95%), second only to Hydrographic Region 7, which concentrates the largest rural contingent among the Hydrographic Regions of the entire state, with 19.40% of drinking water consumption for human consumption in rural areas.

Irrigation is one of the main uses of water resources in Santa Catarina, mainly due to the use of water in rice farming, where the flood method is adopted in most irrigated areas, which results in significant consumption, especially between December and February, when a higher water table is needed for plant growth. Other crops are also irrigated, basically vegetables, where sprinkler irrigation is used as a priority.

[3]According to data from the Santa Catarina Institute of Agricultural Planning and Economics (ICEPA), the Atlantic Strand basins account for practically all the water demand for irrigation in the state, with an estimated 128 million m3 /month, considering that the most critical months are between December and February, due to rice irrigation. Of this total, around 125.5 million m3 /month are for irrigating rice and the rest, around 2.5 million m3 /month, for irrigating other crops. In the Inland Slope, demand for other crops stands out (1.4 million m3/month) against 0.05 million m3/month for rice irrigation.

In addition to the activities mentioned above, the use of water resources in Santa Catarina is also associated with hydroelectric power generation, mineral extraction, tourism and leisure activities, fish farming, navigation and the dilution and removal of liquid effluents.

2.6- Main agricultural activities in the Far West region of Santa Catarina

The agricultural production systems developed in the western region of Santa Catarina are characteristic of diversified family farming that is integrated with the market, more specifically with agro-industry. The family farming system in this region originated from the previous experience of its settlers, the natural resources, the land structure and the family nature of production.

The productive efficiency of this form of agricultural organization in this region is undeniable, since, occupying only about a quarter of Santa Catarina's territory, it contributes more than half of the state's Agricultural Production Value (TESTA et al., 1996).

Santa Catarina's main agricultural products are grown under rainfed conditions, with corn standing out as one of the main crops that provides feed for pig breeding. In addition to corn, soybeans, tobacco, cassava, beans, rice, bananas and potatoes are also grown. The state is also an important producer of garlic, onions, tomatoes, wheat, apples, grapes, oats and barley.

The agricultural area cultivated in the 2010/2011 harvest was approximately 59,110 ha of harvest corn, 11,515 ha of off-season corn, 28,820 ha of soybeans, 2,050 ha of harvest beans, 1,025 ha of off-season beans, 11,580 ha of tobacco and 4,300 ha of wheat (EPAGRI, 2011, unpublished). The main soil management systems practiced on these crops are no-till and minimum tillage, both by mechanical and animal traction.

In the case of the western region of Santa Catarina, it can be seen that in the period 19962002, the area under corn increased from 76,632 hectares to 78,330 hectares, an increase of 11.9% in the region.

It should also be noted that 66% of rural establishments in the region have between 10 and 20 hectares, which shows that small family farming predominates in the region. As a result, the region is nationally known for its enormous productive capacity.

2.7- Soil preparation and management

Soil preparation or management is a practice that acts directly on its structure which, in turn, interacts with or affects a series of profile characteristics, modifying the variables linked to it (VIEIRA, 1985). It consists of simple practices that are indispensable to the proper development of crops and comprises a set of techniques that, when used rationally, provide high productivity, but if misused, can lead to the destruction of soils in the short term and even desertification of large areas (EMBRAPA, 2013).

The crop planting stage is one of the most important in the production process, as it involves the definition, planning and execution of measures and operations that will contribute decisively to obtaining satisfactory yields and the desired return on the capital invested.

The installation of the crop begins with an efficient soil conservation program, aimed at its proper use within technical and economic limits, ensuring rewarding harvests while preserving its productive potential.

Conceptually, conservation practices can be considered as techniques designed to maintain and/or increase the productive capacity of the land, aiming, in addition to controlling erosion, to improve the physical, chemical and biological conditions of the soil (FANCELLI and DOURADO NETO, 2000).

Soil preparation systems can affect soil density, porosity and water storage throughout the profile, directly affecting crop development and productivity (STONE and MOREIRA, 2000).

According to Fancelli and Dourado Neto (2000), soil preparation aims to create favorable conditions for seed germination and plant development, represented by satisfactory aeration, humidity and temperatures, the elimination of physical and chemical impediments, as well as helping to control weeds.

2.7.1 Reduced tillage or minimum tillage

Minimum tillage, as the name implies, consists of minimal soil preparation. Minimum tillage is suitable for non-compacted soils, without the need for liming, phosphating or plastering, or in soils where there are no pests. This system would also be suitable for more sloping areas, where erosion problems are more critical (EMBRAPA, 2013). The aim of minimum tillage is to reduce the number of agricultural operations required to prepare the soil before sowing. In this system there is a reduction in soil compaction simply because the number of machinery movements is reduced. In minimum tillage, the number of passes of machinery and equipment is much smaller than in the conventional system, causing less soil disturbance. Only in the rows, where planting will take place, will the soil be disturbed. The inter-rows remain free, without the movement of machinery and equipment, which means that the soil retains its characteristics.

Minimum tillage, when compared to the traditional tillage system, has some advantages such as:

• Possibility of planting in rainy seasons, which can mean planting can be brought forward by up to a few months;

• More intensive use of the planting area, as the interval between harvesting and replanting is shorter;

• Reduction of erosion;

• Reducing the use of machinery, implements and fuel;

• Weed control.

2.7.2 No-till system

Research into no-till farming practices began at the Rothamsted Experimental Station

(England) in 1940 and in Michigan (USA) around 1946. In Brazil, studies on no-till farming began in the state of Paranà in 1971 and since then, the area under no-till farming has expanded significantly (FANCELLI and DOURADO NETO, 2000).

More than forty years after its introduction in Portugal, no-till farming has become a widely accepted conservation technology among farmers, with systems adapted to different regions and different technological levels, from large to small farmers using animal traction.

This production system requires care in its implementation, but once established, its benefits extend not only to the soil, but also to crop yields and the competitiveness of agricultural systems. Due to the drastic reduction in erosion, it reduces the potential for environmental contamination and gives farmers a greater guarantee of income, as production stability is increased compared to traditional soil management methods. Due to its beneficial effects on the soil's physical, chemical and biological attributes, it can be said that no-till farming is an essential tool for achieving sustainable agricultural systems (EMBRAPA MILHO E SORGO, 2013).

No-till or direct sowing is a system whereby crops are planted in unturned soil and protected by *mulch made from crop* residues, plant coverings sown for this purpose and weeds controlled by combined chemical methods.

It should be noted that, strictly speaking, there is no tillage, but it is restricted to the sowing furrow, in order to ensure adequate contact between the seed and the soil, which is carried out by specialized machinery (FANCELLII and DOURADO NETO, 2000).

As *already* mentioned, from a conservation point of view, no-till farming is one of the most efficient systems for preventing and controlling erosion, which would be enough to justify its adoption. However, other substantial benefits of a different nature resulting from the adoption of this production system are pointed out below, such as:

- Enables crops to be sown at the right times;

- Contributes to reducing fuel consumption in agricultural activities;

- Reduces machine traffic in the area;

- It can help reduce the number of terraces in the field;

- Provides greater moisture conservation in the soil and greater use of available water by plants;

- It contributes considerably to maintaining satisfactory levels of organic matter in the soil;

- It provides a lower temperature range in the soil, favoring the physiology and development of the plant's root system;

- Contributes to improving the total porosity of the soil;

- Provides greater tolerance to periods of drought;

- It ensures a greater chance of obtaining higher yields as it provides better conditions for plant development.

2.7.3 Land use through the crop-livestock integration system

Crop-livestock integration is a technique, also known as rotating annual crops with pastures, in which producers use the land for both animal and plant production, taking turns according to the time of year. Among the benefits found in implementing the technique are: an increase in income, an improvement in the soil's situation with increased efficiency in weed control, increased soil protection against erosion, a reduction in the chemical, physical and biological degradation of the soil in question, making better use of machinery, as well as an increase in the need for labor, generating more local jobs (ALVARENGA and NOCE, 2005).

In crop-livestock integration, when the land is not being used to grow a particular crop, the area becomes pasture and the farmer can dedicate himself to livestock and vice versa. In the case of corn and soybeans, planting takes place in late summer and early spring (September to November) in Santa Catarina and the states above it. Harvesting takes place from January to March or April. In the traditional model, the land would only be used for the next planting (in spring and summer) and the pasture would be dry in winter. In crop-livestock integration, the producer can plant pasture for the cattle right after the harvest. Between July and August, during the dry season, the pasture will be green and able to support a larger number of animals (USP, 2013).

According to the Ministry of Agriculture, Livestock and Supply (MAPA, 2007), in addition to some of the benefits mentioned above, the crop-livestock integration system has other objectives and advantages such as:

• **Produce pasture, forage and grain for animal feed in the dry season** - in addition to producing silage and grain, crop-livestock integration allows the pasture produced in the consortium to be used during the dry season. Correcting the soil improves the development of the root system of the forage, which then goes deeper into the soil and absorbs water at greater depths, keeping it green for longer.

• **Recovering soil fertility with tillage in degraded pasture areas** - chemical soil correction and fertilization for crop cultivation recover soil fertility. This increases the supply of nutrients to the pasture and, consequently, its production potential.

• **Reducing the costs of both agriculture and livestock** - the crop-livestock integration system increases the productivity of crops and pastures, reduces the consumption of pesticides and rationalizes the use of labor. As a result, production costs are reduced.

• **Improving the soil's physical and biological conditions with grazing in arable areas** - grazing leaves considerable amounts of straw and roots in the soil. This increases the amount of organic matter, which is essential for improving the soil's physical structure. It is also a source of nutrients for soil organisms. This new environment created in the soil by crop-livestock integration is fundamental to increasing the productivity of both crops and livestock.

2.8- The use of software for hydroagricultural modeling

Computer software used in irrigation management are programs for microcomputers that calculate crop and irrigation water requirements based on climate, soil and crop data. In addition, the programs make it possible to establish irrigation schedules for different management conditions, and calculate the water supply scheme of a project for different cropping patterns (PEREIRA, 2004).

There are several computer models for determining irrigation schedules, the first of which was developed by Jensen (1969) using climate, crop and soil data (VIANA, 1997).

Smith (1989 apud GUIMARÃES 1993), with the aim of obtaining practical criteria for irrigation schedules and water distribution, developed a simulation model for programming irrigation management called "CROPWAT". This FAO model is based on the relationships between water, soil, plant and climate and allows situations to be simulated in such a way as to present the best options for efficient water use and crop yield reduction.

Ribeiro (1992) developed the CADIR computer model in order to draw up irrigation calendars for the maize crop in the Curu-Pentecoste irrigated perimeter, achieving satisfactory results. This model comprises four modules: crop, soil, climate and water balance. It also allows you to manipulate the data required to determine the irrigation calendar.

Other computer software used in irrigation management is the SIMDualKc model, developed by Godinho (2007) at the Center for Rural Engineering Studies at the Instituto Superior de Agronomia of the Technical University of Lisbon, with a view to calculating crop evapotranspiration (ETc) and scheduling irrigation, using the crop coefficient methodology proposed by Allen et al. (1998). This method considers soil evaporation and crop transpiration separately, analyzing how water from precipitation and irrigation is used by crops. The main aim of the SIMDualKc model is to develop options for scheduling irrigation, particularly for crops with partial soil cover, especially vegetables and orchards, and/or for systems with high-frequency irrigation, such as the microgreen.

According to Pereira (2003), the advantage for irrigators of being given guidance on irrigation management is to have a guide that allows them to know approximately what consumption they are expecting from the crop they are growing and, therefore, to establish their own irrigation schedule, depending on the system they use on their property.

The determination of irrigation needs and the definition of allocations can be facilitated by the use of water balance simulation models. However, they require adequate parameterization with regard to the soil and the crop and, for irrigation management, ways of getting the information to farmers, including the support of the web (PEREIRA, 2007).

2.8.1 Simulating the water balance for crops using the ISAREG model

Deficit irrigation, in which the crop's need for water is only partially met, can be an appropriate technology or lead to economically unviable production losses. Over-irrigation leads to wasted water, higher production costs and inappropriate management of available water resources. In this sense, the adoption of appropriate irrigation technologies - when and how much to irrigate - is of fundamental importance. To this end, among other technologies, it is possible to simulate the crop's water balance using models and then carry out an economic analysis using the results of this simulation.

The ISAREG model (TEIXEIRA and PEREIRA, 1992) calculates the water balance in the soil according to the methodology described by Doorenbos and Pruitt (1977) and Allen et al., (2000).

ISAREG is a water balance simulation model designed to program irrigation for a given soil-climate-crop combination, as well as to evaluate water consumption by crops and the irrigation schedules adopted (TEIXEIRA and PEREIRA, 1992; LIU et al., 1998). It uses the methodology for calculating crop evapotranspiration proposed by Allen et al. (1998), and the calculations can be carried out at daily, decadal or monthly intervals.

This model can be used for different irrigation options such as: evaluating irrigation for maximum production, simulating an irrigation scheme using selected limits, evaluating an irrigation scheme when water is applied on specific dates, looking for an optimal irrigation scheme under conditions of limited water availability, with constant or variable endowments, and also running the water balance without irrigation and calculating irrigation needs.

The ISAREG model database is organized into "meteorological" and "agronomic" files, which include data on: effective precipitation (Pe, mm), reference evapotranspiration (ETo, mm); crop data (crop cycle phases and respective crop coefficients (Kc); root system depth (z, mm); and the fraction of soil water available without causing water deficiency ("p" index, decimal); soil data referring to the multiple soil layers (Ds); water content relative to field capacity and permanent wilting point; data relative to capillary rise and deep percolation, which can be calculated using parametric equations as a function of the soil's hydric properties, the depth of the water table, ETc and the water available in the soil (LIU et al., 2006); data on irrigation options, where it is possible to save the various simulation options evaluated; data on water restrictions, referring to the different simulated irrigation schemes in terms of the volumes of water available and the periods in which they are available.

After verifying whether or not irrigation is necessary, the ISAREG model makes it possible to program it by calculating the day (when) and the volume of water to be applied (how much) for each irrigation; it also makes it possible to estimate the loss of production if the crop eventually suffers water stress, thus determining the theoretical irrigation needs, regardless of how it is carried out, provided that the crop in question is adequately supplied with water. The ISAREG model also makes it possible to evaluate a given irrigation schedule and define its parameters, i.e. calculate annual irrigation needs and flow rates, by constructing statistical series of these parameters.

With these analyses, it is possible to assess probable losses in the crop's agricultural productivity during periods when there are water restrictions on the plant's consumption, as well as to quantify and assess the need to meet the crop's water demand through irrigation systems.

Therefore, after simulating various combinations of management, climate, soil and crop, the ISAREG model makes it possible to assess the water demands on crops and even whether the region under study has the water capacity to supply this irrigation activity, making it possible to make strategic changes to water management in order to optimize its use, both for private crops and for the region as a whole.

2.9-- Regional climate modeling using quantitative precipitation analysis

Water scarcity in the future, as already mentioned, could become the reason for major global conflicts between nations, and could bring with it serious damage to the availability of this commodity for human consumption, as well as for agriculture. Furthermore, according to Nobre and Assad (2005), the global average surface temperature of the planet has been increasing over the last 120 years, having already reached +0.6 to 0.7 °C. Most of this warming has occurred in the last 50 years, with the last decade presenting the three warmest years in the last 1000 years of the Earth's recent history. Recently, the IPCC report (2007) reinforced these facts, stating that it is very likely that this increase in air temperature is the result of anthropogenic actions. According to Nobre and Assad (2005), the increase in temperature leads to greater evapotranspiration, reducing the amount of water in the soil, even if rainfall does not decrease significantly.

Concerned about such future climate scenarios, researchers such as Marcelino et al. (2004), Severo (1994), among others, have carried out studies with the aim of analyzing the behavior of rainfall over time.

Meteorological research institutions such as: the National Institute for Space Research (INPE), through the Research Group on Climate Change (GPMC), the National Institute of Meteorology (INMET), the Institute of Astronomy, Geophysics and Atmospheric Sciences of the University of São Paulo (IAG), the Brazilian Foundation for Sustainable Development (FBDS), among others, are also evaluating and developing future climate projection models in order to support farmers and other decision-makers in formulating policies on the impact of climate change, vulnerability and adaptation measures in various fields.

Climatological institutions have developed some climate models such as: HadCM3 (British Meteorological Service), Eta-HadCM3 (University of Belgrade together with the Institute of Hydrometeorology of Lugoslavia), CGCM3.1.T47 (Canada), CCSM3 (USA), among others, which are ocean-atmosphere coupling algorithms, numerical tools whose purpose is to analyze climate variability and change over a long period of time and which allow future climate projections for various locations.

Through the use of these climate models, it becomes possible to develop studies that indicate the trend of future events and also subtle changes that occur continuously over several decades or centuries, such as the scarcity or excess of rainfall in a given region, making it possible to assess whether rainfall will remain frequent or whether there will be periods of long droughts.

However, global climate models (GCMs) have a very coarse spatial resolution, which does not capture changes at smaller scales, such as topographical and land cover characteristics. It is therefore necessary to refine this grid using downscaling techniques. One method widely used in climate modeling is to refine the grid using regional climate models (RCM) using the MCG conditions (CHOU et al., 2011).

Climate projections from RCMs are useful in studying the impacts of climate change, given their ability to better understand local surface features (CHOU et al., 2011).

The Center for Weather Forecasting and Climate Studies (CPTEC) of the National Institute for Space Research (INPE) has been successfully using the Eta regional climate model operationally (24 hours a day) since 1996 to generate weather forecasts for South America.

Chou et al. (2005) showed that the regional Eta model presented better results when compared to the general circulation model (global scale) MCG-CPTEC for seasonal climate forecasting purposes. One of the most widely used global climate models for generating boundary conditions is the HadAM3P model from the Hadley Center (UK), which is the atmospheric version of the HadCM3 coupled ocean-atmosphere model from the same center. Pesquero et al. (2010) showed better results from the Eta regional climate model compared directly with HadAM3P precipitation data from the present climate (19611990).

In view of this, this study used the CPTEC Eta regional model with the boundary conditions of the HadCM3 global model (CHOU et al., 2011, MARENGO et al., 2011), thus being called the Eta-HadCM3 model - whose simulations for the present (1961-1990) and future (2010-2099) were kindly provided by INPE's Center for Earth System Sciences (CCST). In comparison with climatologies (1961-1990) based on reanalyses over South America, the HadCM3 model generally showed better results than HadAM3 (PESQUERO, 2009).

The Eta model is so named because it uses the vertical coordinate η, which is suitable for regions with steep orography such as the Andes (MESINGER, 1984). The prognostic variables are temperature, specific humidity, horizontal wind, surface pressure, turbulent kinetic energy and the amount of water and ice in the clouds. The precipitation model is based on the Betts-Muller-Janjic cumulus parameterization scheme (Janjic, 1984) and the Zhao cloud microphysics scheme (ZHAO et al., 1997). The soil-surface scheme (CHEN et al., 1997, EK et al., 2003) is based on temperature and humidity at 4 depths: 10 cm, 30 cm, 60 cm and 100 cm. The scheme distinguishes 12 types of vegetation and 7 types of soil texture.

The seasonal climate version of the Eta model was adapted for decadal integrations in order to analyze climate change scenarios related to different levels of CO_2 concentration in the atmosphere (CHOU et al., 2011). Some changes were made to the original Eta-CPTEC code so that CO_2 concentrations could vary according to HadCM3. A linear interpolation was made on the time scale to avoid annual CO_2 data being generated with sudden jumps (MARENGO et al., 2011). For the simulation of the present climate (1961-1990), the CO_2 concentration was set as constant at 330 ppm (parts per iiiilliào). The HadCM3 boundary conditions were updated every 6 hours; sea surface temperature values were obtained from the coupled ocean-atmosphere model, and Iinear interpolations were also applied to the vegetation index, which was prescribed as monthly averages.

The integration of the model began on 01/01/1960, with one year for the model to reach equilibrium, i.e. during this period the data generated by the model was not considered reliable due to initial variations, so the simulation of the present time comprises the period from 1961 to 1990, i.e. a continuous integration of 31 years. Initial soil moisture data started with the January climatology and albedo from the seasonal climatology. The Eta-HadCM3 model was set up to run with a horizontal resolution of 40 km, 38 vertical levels and a 90 s time step.

2.9.1 Climate modeling for future scenarios

Among the factors that influence the behavior of future scenarios to be estimated by climate models is the emission rate of greenhouse gases (especially CO_2). It is common knowledge that the long-wave absorption capacity of greenhouse gases leads to a warming of the Earth's system. Two questions are always pertinent: how responsible are anthropogenic actions for the rate of global warming; and how accurate are climate models in predicting this rate. With this in mind, the Intergovernmental Panel on Climate Change (IPCC) of the United Nations (UN) was set up to study the subject using different methodologies and technologies (climate modeling). Figure 11 below summarizes the behaviour of climate modelling in representing the last century in relation to the temperature anomaly considering anthropogenic action.

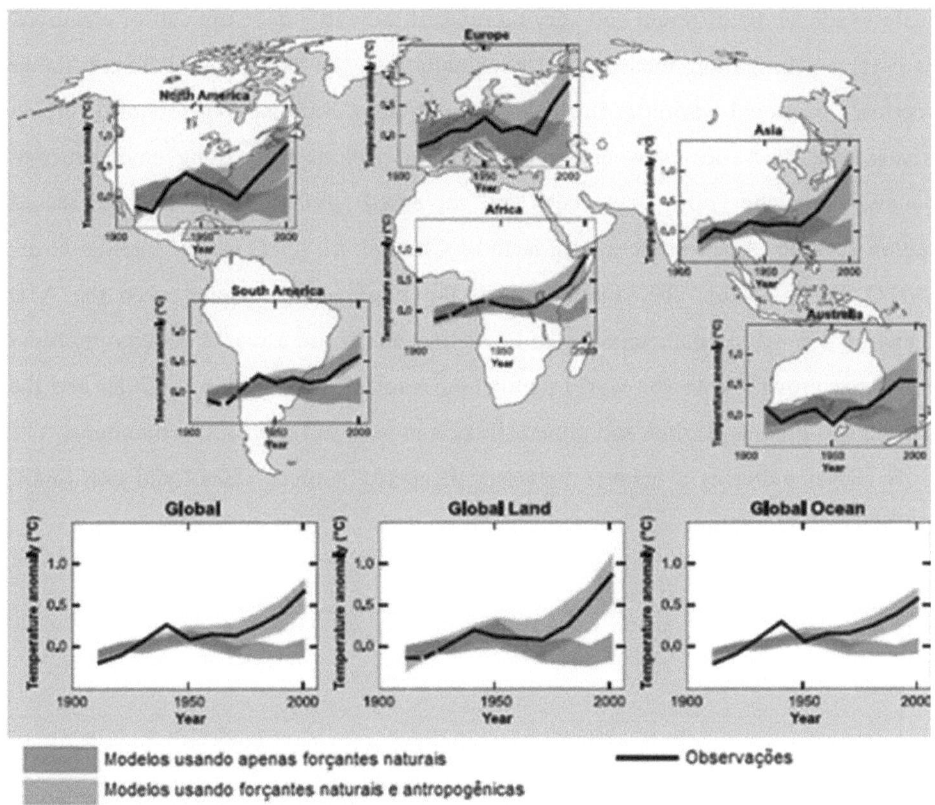

Figure 7 - Comparison between changes in surface temperature on a global scale and those simulated by climate models using natural forcings alone or natural and anthropogenic forcings. Source: IPCC (2007).

Figure 7 shows decadal averages from 1906-2005 of the observations (black lines) of anomalies (in relation to the climatological average 1901-1950). The lines are dashed where the spatial coverage of observations was less than 50%. Blue bands show the result of 19 simulations of 5 climate models using only natural forcings such as solar and volcanic activity. Red bands represent the result of 58 simulations of 14 models using natural and anthropogenic forcings (greenhouse gas emissions due to industrialization).

The Special Report on Emissions Scenarios (SRES), published by the IPCC in 2000, projected an increase in global greenhouse gas emissions of more than 25% between 2000 and 2030. However, given the dynamics in which emissions occur on a global scale - such as the different and varying rates of industrial development of countries, as well as mitigation measures - the simulations are based on different future scenarios of global warming. Basically, the scenarios combine two sets of divergent trends: one set varies between strong economic values and strong environmental values, and the other set between increased globalization and increased regionalization. As in other works, such as Chou et al., (2011) and Marengo et al., (2011), in this work, the simulations of the Eta-HadCM3 model used the A1B scenario. The main characteristics of the A1 scenarios are a world undergoing rapid economic growth, with the world population reaching nine billion by 2050 and the consequent gradual decline and rapid diffusion of new and efficient technologies. The A1B subset indicates a balance between all energy sources (fossil and non-fossil) (PESQUERO, 2009).

3. MATERIAL AND METHODS

This study evaluated two different scenarios:

1st scenario - present: evaluation of the values of net irrigation water to be applied to the corn crop from 1988 to 2010 under three soil management systems (minimum tillage, no-till and perennial pasture) and in three typical soils of the Far West region of the State of Santa Catarina (Cambissolo, Latossolo and Nitossolo).

For this evaluation, we used climatological data on maximum (Tmax) and minimum (Tmin, °C) temperatures, relative humidity (RH, %), wind speed (Vv, ms-1) and insolation (In, Wm-2) for the city of Sao Miguel do Oeste, provided by the Santa Catarina Center for Environmental Information and Resources and Hydrometeorology (EPAGRI/CIRAM).

2nd scenario - Future: using estimates of future climate data generated by the Eta-HadCM3 climatological model made available by the National Institute for Space Research (INPE), estimates were made of net blades for future scenarios (2014-2036), based on data on maximum temperature (°C), minimum temperature (°C), relative humidity (%), wind speed (m s-1) and insolation (W m-2).

Based on the results generated in the two scenarios, an evaluation and comparison was made between the irrigation rates for the different soil types and the three tillage systems. In addition, it was analyzed whether, for the future scenario (2014-2036), there would be a need to increase the Liquid Lamina (LL) values to be applied to the corn crop under the conditions studied for the region when compared to the present scenario (1988-2010).

3.1 - The Western Region of Santa Catarina and its main characteristics

3.1.1 - Hydrographic Regions of the State of Santa Catarina

The state of Santa Catarina is located at geographic coordinates 25°57'41"S and 29°23'55"S and 48°19'37"W and 53°50W' W. It has a total area of 95,346.181 km^2 , which represents 1.12% of the national surface and 16.61% of the southern region of the country. The state is located in the southern region of Brazil and shares state borders with the states of Paranà (to the north) and Rio Grande do Sul (to the south). It also shares a border with Argentina (to the west) and is bordered to the east by the Atlantic Ocean (Figure 8). The state of Santa Catarina has the city of Florianópolis as its administrative capital and has 293 municipalities that make up the Federation of Municipalities of Santa Catarina - FECAM.

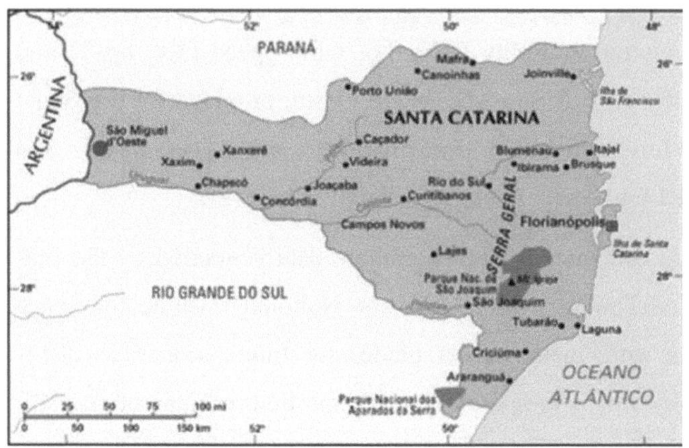

Figure 8 - Image showing the state of Santa Catarina and its borders with other states, the country and the ocean.

Source: EPAGRI, 2011.

With regard to Hydrographic Regions, according to the current division adopted by the National Water Agency (ANA, 2005) and the Government of the State of Santa Catarina (PRAPEM, 2006), the rivers draining the state territory of Santa Catarina are part of three major Hydrographic Regions: the Paranà Hydrographic Region, the Uruguay Hydrographic Region and the South Atlantic Hydrographic Region.

Santa Catarina's hydrographic network has the Serra Geral as the main watershed that forms the two independent drainage systems in the state: the integrated system of the *Interior Strand*, comprising 11 basins that make up the Paranà-Uruguai basin, which refers to the rivers in the western half of the state. The *Atlantic Strand* system, made up of a group of 12 isolated basins that flow eastwards directly into the Atlantic, refers to the rivers in the eastern half of the state of Santa Catarina. For the purposes of managing water resources, the state has been subdivided into 10 Hydrographic Regions (HR). The basins of the Interior Strand are part of five Hydrographic Regions: 1- Far West, 2- Mid West, 3- Vale do Rio do Peixe, 4- Lages Plateau and 5- Canoinhas Plateau. The other Hydrographic Regions are part of the Atlantic Strand: 6- Baixada Norte, 7- Vale do Itajai, 8- Litoral Centro, 9- Sul Catarinense and 10- Extremo Sul Catarinense (Figure 9).

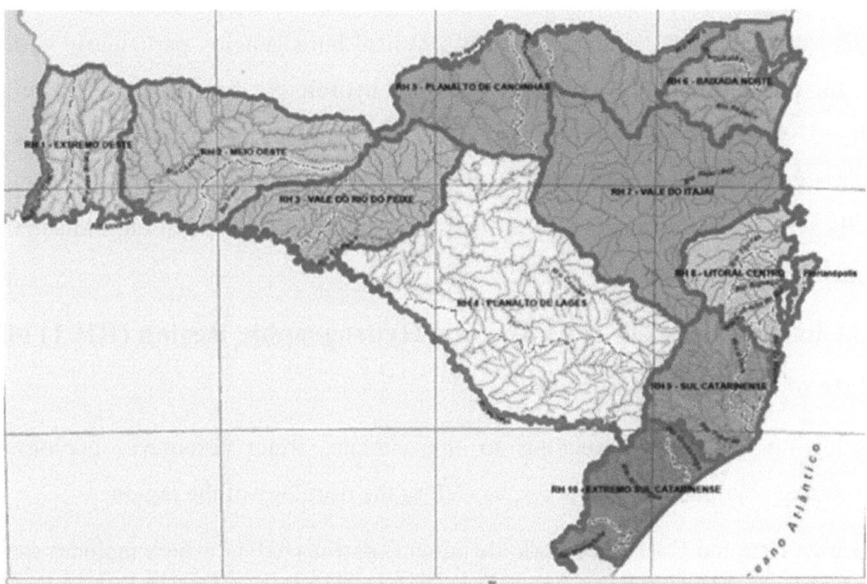

Figure 9 - Hydrographic regions in the state of Santa Catarina. Source: EPAGRI, 2011.

The Inland Slope covers a total of 58,784 km^2 , corresponding to around 62% of the state's territory. The Atlantic Slope extends over an area of approximately 36,284 km2 and corresponds to 38% of the state of Santa Catarina.

It is worth noting that the boundaries of the Hydrographic Regions are defined on the basis of a number of criteria:

• The watershed should be the basic unit for planning the use, conservation and recovery of natural resources;

• The river basins that make up each Hydrographic Region (HR) must have

51

homogeneous physical and socio-economic aspects;

• The geographical area of the different Hydrographic Regions (HR) must maintain a certain degree of identity with that of the existing municipal associations; and

• The number of municipalities located in each Hydrographic Region should not be too high and the maximum area of each region should not be too large.

In view of this, the Hydrographic Regions have a considerable degree of geographical coincidence, a high degree of physical homogeneity, particularly with regard to geomorphology, geology, regional hydrology, type of soil relief, agricultural suitability and current land use. The high degree of socio-economic homogeneity of the basins that make up the same region is also taken into account, especially with regard to the size of properties, type of rural exploitation, industrial activity, among other aspects.

3.1. 2- Characteristics of the Far West Hydrographic Region (RH 1) of the State of Santa Catarina

The following are aspects relating to the climate, water resources, geology, geomorphology and mineral resources, as well as the pedology of the region

Hidrogràfica Extremo Oeste do Estado de Santa Catarina (RH 1), which includes the city of Sao Miguel do Oeste, located in the center of this Hidrogràfica Region (RH 1) (Figure 10) and for which this study is intended.

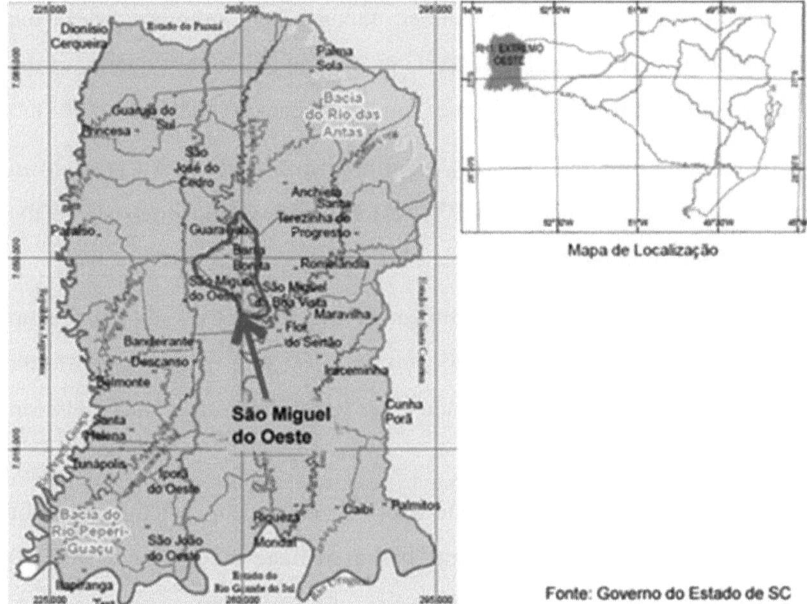

Figure 10- Extremo Oeste Hydrographic Region (RHl) in the state of Santa Catarina and indication of the location of the city of Sao Miguel do Oeste within it.

According to Epagri (2002), based on Koppen's climatic characterization, RH 1 lies at the interface of two climatic types, "Cfa", a humid mesothermal subtropical climate with hot summers, and "Cfb", a humid mesothermal temperate climate with mild summers, with minimum temperatures of less than 15°C, lasting around 120 days during the months of May to September. Frosts and the penetration of polar fronts are frequent during this period.

The subtropical mesothermal humid climate type "Cfa" is distributed over 96% of the area, corresponding to 5,603 km^2 and the temperate mesothermal humid climate type "Cfb", over 3%, i.e. 190 km2, occurring in the northern portion, near the municipality of Campo Erê.

In terms of temperature, there is an increasing gradient from north to south, and from the highest to the lowest altitudes. The average annual temperature in this hydrographic region ranges from 16°C (north) to 20°C (south), where the average maximum temperature ranges from 24°C to 28°C in the hottest quarter of the year, which corresponds to the months of December, January and February, while the average minimum temperature ranges from 11°C to 15°C in the coldest quarter of the year, which corresponds to the months of June, July and August (FEHIDRO, 2008).

As for the average annual relative humidity, there is also a north-south gradient in RH 1, with values of between 74% and 82%, which occurs homogeneously in the Antas and Peperi-Guaçu river basins.

The behavior of total annual rainfall is similar in the two basins that make up Hydrographic Region 1, ranging from 1,700 mm to 2,100 mm, with a south-north gradient. In the northwest of the Peperi-Guaçu river basin there are values between 2,100 and 2,300 mm

Specifically, the city of Sao Miguel do Oeste has an average annual temperature of 18.8°C. With regard to annual rainfall, this city has an average accumulation of 2250 mm of rain (EMBRAPA, 1998).

With regard to insolation, the Far West Hydrographic Region (RH 1) has annual insolation values of between 2,200 hours/year and 2,400 hours/year.

Evapotranspiration, which is strongly influenced by average air temperature, the degree of air saturation (relative humidity) and wind speed, has annual values of around 900 to 1,000 mm/year in practically the entire RH 1 area.

3.1.3 - Characterization of surface and underground water resources in the far west of Santa Catarina

With 5,834.92 km^2 , the Far West Hydrographic Region (RH 1) is made up of the basins of the Peperi-Guaçu and das Antas rivers, which rise on the border between the states of Paranà and Santa Catarina and flow into the right bank of the Uruguay River. The Peperi-Guaçu River borders Argentina for a length of approximately 250 km. Among the tributaries on the left bank, located in Santa Catarina, are the rivers das Flores, Maria Preta and Uniao. The 193 km long Antas River drains an area of 3,651.70 km2 along with other direct tributaries of the Uruguay River. Its main tributaries, the Sargento and Capetinga rivers, are located on the left bank.

From a hydrogeological point of view, 99.41% of the area corresponding to RH 1 falls within the Serra Geral Hydrogeological Province, on the Serra Geral Aquifer Unit, corresponding to a surface area of 5,800.39 km2.

In terms of basins, the Serra Geral Aquifer Unit covers 99.34% of the area of the Antas river basin (3,627.56 km2) and 99.52% (2,172.82 km2) of the Peperi-Guaçu river basin.

3.1.4 - Pedological Characterization of the Hydrographic Region of the Far West of Santa Catarina

As can be seen in Table 4 and Figure 10, there are four mapping units in RH 1: Cambissolo Hàplico, Latossolo Bruno, Latossolo Vermelho and Nitossolo Vermelho, whose areas correspond to, respectively, 3,782 km^2 , 403.89 km^2 , 323.39 km^2 and 1,261.58 km^2 . Therefore, the Cambissolo Hàplico unit predominates (64.82%), followed by the Nitossolo Vermelho unit (21.62%).

Table 4- Geological mapping of the Extremo Oeste/SC Hydrographic Region.

Mapping Units	River das Antas		Peperi- Guaçu River Basin		Region Hydrographic 1	
	Area (Km)^2	%	Area (Km2)	%	Area (Km2)	%
Hepatic Cambisol	2.374,96	65,04	1.407,04	64,45	3.782	64,82
Bruno latosol	385,32	10,55	18,57	0,85	403,89	6,92
Red Latosol	183,16	5,02	140,23	6,42	323,39	5,54
Red Nitosol	667,90	18,29	593,68	27,19	1.261,58	21,62
Urban Area	4,11	0,11	5,82	0,27	9,92	0,17
Mass of water	34,4	0,94	10,84	0,50	45,29	0,78
No Information	1,81	0,05	7,03	0,32	8,84	0,15
Total	**3.651,70**	**100**	**2.183,22**	**100**	**5.834,92**	**100**

Source: EMBRAPA (1999).

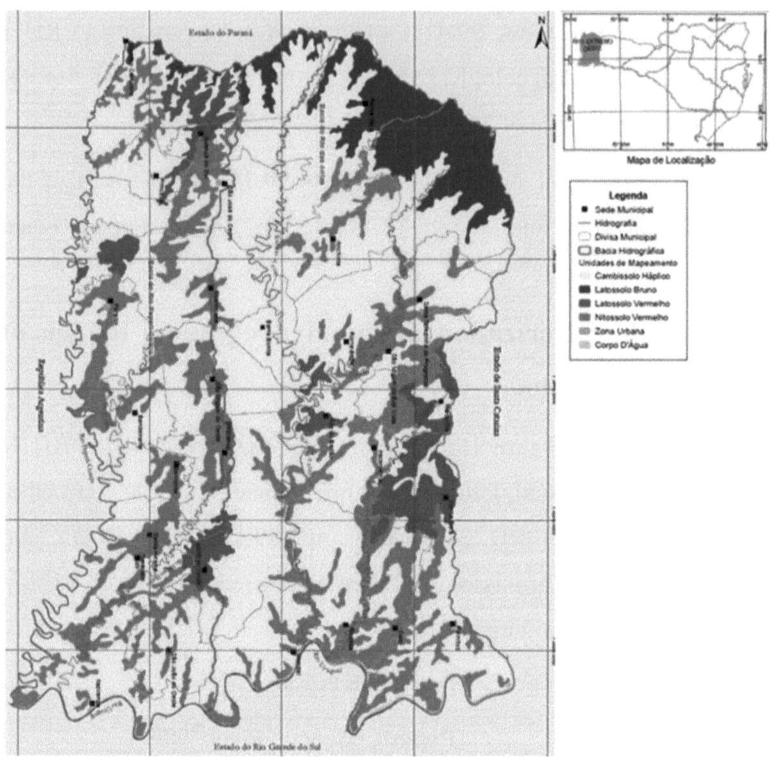

Figure 11 - Pedological map of the Extremo Oeste/SC Hydrographic Region. Source: Government of the State of Santa Catarina.

Analyzing the data by hydrographic basin, shown in Table 4 above and Figure 11, it can be seen that 65.04% of the area of the Antas river basin is Cambissolo Hàplico, covering an area of 2,374.96 km^2 .

Another unit that stands out in this basin is the Red Nitosol, which covers an area of 667.90 km2, or 18.29% of the basin's total area. The Latossolo Bruno and Latossolo Vermelho units occupy less significant areas, 385.32 km2 and 183.16 km2, respectively.

In the Peperi-Guaçu river basin, 64.45% of its area is mapped as Cambissolo Hàplico, corresponding to an area of 1,407.04 km^2 and 27.19% is mapped as Nitossolo Vermelho, covering an area of 593.68 km2. Red Latosol is present in 6.42% of the area, or 140.23 km2.

3.1.4.1 - Characteristics of the soils found in the Far West of Santa Catarina

According to the pedological map above (Figure 10), it can be seen that the predominant soils in the far west of Santa Catarina are cambisols, nitosols and latosols. Some of their characteristics are described below:

•   **Cambissolo**: according to Embrapa (2013), this soil is characterized by being strongly, imperfectly, drained, shallow to deep and poorly developed, with an incipient B horizon (poorly evolved, with the presence of rock fragments and primary minerals, poor development of structure and color), poorly advanced pedogenesis and relatively higher silt contents at depth. The incipient B horizon occurs below the surface horizon of any type, including the chernozemic A horizon (dark-colored surface horizon, rich in organic matter and very fertile). Sometimes the clay content in the subsurface horizon can be lower than the horizons above.

•   **Latosols**: soils with a latosolic B horizon (subsurface horizon, uniform in color, texture and structure), very advanced evolution with severe weathering. The intense weathering of the mineral constituents results in a relative concentration of resistant clay minerals (iron and aluminum oxides and hydroxides). Inexpressive mobilization or migration of clay at depth. They are deep soils, generally very poor, occupying the oldest and most stable surfaces of the landscape. They have a variable texture, on average very clayey, porous, with high permeability and very low natural fertility.

•   **Nitosols:** soils with a clayey or very clayey texture with no significant increase in clay content from the surface to the subsoil. In short, they are soils made up of mineral material with 350 g/Kg or more of clay, non-hydromorphic and with a subsurface horizon immediately below the A horizon, with little textural differentiation and a well-developed blocky structure and very clear cerosity. In general, they are deep, well-drained, predominantly reddish or burnished, moderately acidic and of very variable natural fertility. It is formed on basic rocks and generally occupies the middle and lower portions of undulating to strongly undulating slopes, so, being on slopes that have been moved, it is at high risk of erosion.

3.1.4.2   - Characteristics of the city of Sao Miguel do Oeste

The city of Sao Miguel do Oeste is located in the far west of the state of Santa Catarina, at 26° 72' latitude (S) and 53° 51' longitude (W) (Figure 12).

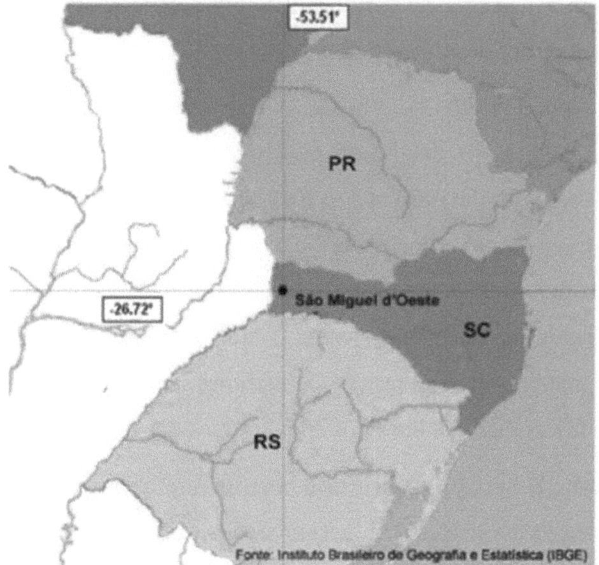

Figure 12 - Geographical coordinates of the municipality of Sao Miguel do Oeste/SC

As far as municipal boundaries are concerned, Sao Miguel do Oeste is located:

to the south with the municipality of Descanso; to the east with the municipalities of Barra Bonita, Romelândia and Flor do Sertao; to the west with the municipalities of Bandeirante and Paraiso; and to the north with the municipality of Guaraciaba.

The climate, according to the Koeppen classification, is of the "Cfb" type (temperate wet mesothermal climate with mild summers).

This region has an average annual temperature of 18.8°C and an average annual rainfall of 2250 mm (EMBRAPA, 1998).

The relative humidity varies from 75% in the driest periods to 82% in the wettest.

The typical soils of this region, as highlighted above, are characterized as: Cambissolo Hàplico, Latossolo Bruno, Latossolo Vermelho and Nitossolo Vermelho

(EMBRAPA, 1999).

3.2 - Collecting soil samples

In order to feed the "ISAREG" model with physical data characteristic of Santa Catarina's soils, representative samples were collected in the far west of the state, and evaluated at the Soil Physics Laboratory of EPAGRI's Experimental Station in Campos Novos/SC.

Soil samples for physical analysis were collected according to the methodology described in (VEIGA, 2011). The definition of the collection points for these samples was pre-determined based on the soil survey of Santa Catarina (EMBRAPA, 2004), through which the location of the occurrence of the three soil classes considered (Cambissolos, Nitossolos and Latossolos) was obtained.

Four samples of each soil class were collected from different properties in each type of management considered, which constituted the repetitions of the sampling units, totaling 144 samples (3 soils x 3 managements x 4 depths x 4 repetitions). The four collection sites containing the three management systems considered (Direct Planting - PD, Minimum Cultivation - CM and Perennial Pasture - PP) were defined using information obtained from the Extension teams at EPAGRI's Municipal Offices and constituted four replications of each treatment (soil class associated with a management system). Samples for physical analysis were taken at four depths (0-10 cm; 10-20 cm; 20-40 cm and 40 to 60 cm).

In order to carry out the physical analyses (retention curve, porosity, density, penetration resistance and soil granulometry) it was necessary to collect just one sample per layer in each plot. This sample was collected using a volumetric ring with internal dimensions of approximately 5 cm in diameter and 4 cm in height, sampling a cylinder of soil with these dimensions, making up approximately 100 cm^3 . This collection was carried out in the center of each layer, i.e. 3-7, 13-17, 28-32 and 52-62 cm deep.

Soil *water* retention curves were drawn up for the three main soil types (Cambissolos,

Nitossolos and Latossolos) and for each of the three predominant management systems (PD, CM and PP) in order to determine the field capacity (CC) and the permanent wilting point (PMP).

The retention curve was determined at the Soil Physics Laboratory of the EPAGRI Experimental Station in Campos Novos, at tensions of 0.2; 0.6; 2; 6 bar (20, 60, 200, 600kPa) using the Richards pressure chamber.

Deformed soil samples for analysis of organic matter content were taken at two depths (0-20 cm and 20 to 60 cm). The samples were collected in the same places as the physical analyses and were analyzed according to the methodology described in Tedesco et al. (2004).

Table 5 below shows the particle size results obtained from the physical analyses carried out on the soil samples.

Table 5 - Soil granulometry and treatments

Soil	Preparation System	Depth (m)	Gi Sand'anulométr(g kg^{-1}) fraction	Silt	Clay
		0,0 - 0,10	475.3	387.6	137.0
	*CM	0,10 - 0,20	504.6	354.3	141.1
		0,20 - 0,40	499.9	360.6	139.4
		0,40 - 0,60	527.1	312.2	160.7
		0,0 - 0,10	336.5	439.7	223.8
Cambisol	*PP	0,10 - 0,20	379.0	397.4	223.6
		0,20 - 0,40	361.6	371.9	266.5
		0,40 - 0,60	451.9	308.3	239.7
		0,0 - 0,10	461.5	375.3	163.3
	*PD	0,10 - 0,20	405.9	371.3	222.7
		0,20 - 0,40	383.7	344.7	271.5
		0,40 - 0,60	407.4	298.1	294.5
		0,0 - 0,10	160.7	342.9	496.4
	*CM	0,10 - 0,20	140.7	311.2	548.1
		0,20 - 0,40	85.0	242.0	673.0
		0,40 - 0,60	121.6	214.0	664.4
		0,0 - 0,10	197.7	436.3	365.9
Nitossolo	*PP	0,10 - 0,20	224.7	362.9	412.3
		0,20 - 0,40	205.6	310.3	484.1
		0,40 - 0,60	199.7	234.5	565.8
		0,0 - 0,10	153.3	357.6	489.1
	*PD	0,10 - 0,20	126.0	281.1	592.9
		0,20 - 0,40	89.0	230.0	681.0
		0,40 - 0,60	79.1	223.3	697.6
		0,0 - 0,10	123.6	441.0	435.4

		0,10 - 0,20	103.5	357.2	539.3
	*CM	0,20 - 0,40	88.7	312.5	598.8
		0,40 - 0,60	74.4	249.6	676.0
		0,0 - 0,10	189.6	470.8	339.5
Latosol	*PP	0,10 - 0,20	165.7	381.2	540.1
		0,20 - 0,40	123.4	276.5	600.1
		0,40 - 0,60	108.2	234.1	657.7
		0,0 - 0,10	137.7	414.5	447.8
	*PD	0,10 - 0,20	75.9	289.6	634.5
		0,20 - 0,40	58.7	275.2	666.1
		0,40 - 0,60	55.6	225.3	719.0

*CM = Minimum Cultivation; *PP = Perennial Pasture; *PD = Direct Planting

3.2.1 - Soil water balance using the "ISAREG" model

Various irrigation management strategies for irrigated crops have been developed using models that simulate the water balance, with the aim of improving the use of water by crops, avoiding water losses through irrigation and improving the performance of irrigation systems (PEREIRA et al., 2003). Water balance models are considered important tools to ensure the best use of water, as all the elements of the balance can be evaluated, and long-term evaluations can be easily developed (DROOGERS et al., 2000). Ji et al. (2007) consider that simulation models play an important role in calculating the water balance, as the physical processes involved are known, and can be used to predict the impact of irrigation on available water resources as well as helping to assess its viability.

According to Pereira et al. (1995), the models can essentially be of two types: water flow simulation models, when the water balance is computed using inflows and outflows and the retention of water in the soil in the root zone; and volumetric soil water balance simulation models, when the water balance is obtained by simulating the volumes of water entering and leaving the soil reservoir at a predetermined time interval.

Pereira et al. (1995) consider that volumetric water balance models are easily parameterized and only require essential soil water characterization and basic crop data, in addition to adopting simplified water-yield functions to evaluate the effects of water deficits on yield reductions.

According to Paz et al. (1996), in ISAREG (Figure 13), which is a water balance simulation model developed by Teixeira and Pereira (1992) at the Instituto Superior de Agronomia (ISA) in Portugal, the water balance is based on the method proposed by Doorenbos and Pruitt (1980) and Doorenbos and Kassan (1980), requiring meteorological, edaphic and agronomic data.

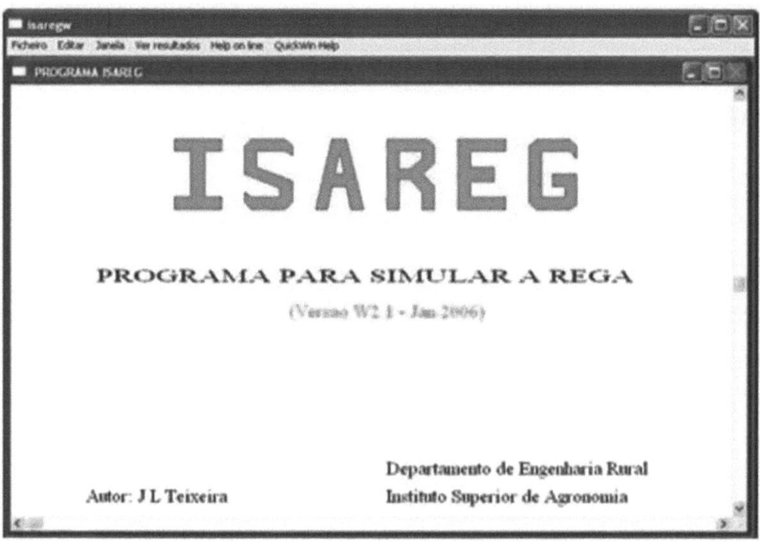

Figure 13 - Initial screen of the program used to quantify the water balance "ISAREG".

In this model, the water balance equation considers a square prism as the horizontal unit of area, whose height is adjusted to root growth. Thus, the volume of soil expresses the available water. In addition, the simulation of the soil profile is developed in multi-layers, allowing a reasonably accurate presentation of the soil compartments of the system.

In the latest version of the ISAREG model, it is combined with two other programs: EVAP56, which calculates reference evapotranspiration (ETo) using the Penman-Monteith method, and KCISA, which calculates the crop parameters required by the FAO methodology (ALLEN et al., 1998). Simulations can be carried out for daily, decadal or monthly periods. In the present work, the simulations were carried out for decadal periods, but for the statistical analyses, the total sum of the net slashes (LL) to be applied to the soil during the corn growing period was used, which was established from September 15 (corn sowing date in the Sao Miguel do Oeste region) to January 24 of the following year (corn harvest date in the region).

It was decided to work with total liquid yields in order to eliminate the zeros present in some of the decadal results. The decadal values of the net blade for the three types of soil and three tillage systems are presented in the Appendix of this work.

The ISAREG model simulates the soil water balance for the total depth of the effective root zone using equation 16 below:

$$\theta_i = \theta_{i-1} + \frac{(P_i - Q_{ri}) + I_{ni} - ET_{ci} - DP_i + GW_i}{1000 z_{ri}}$$

(Equation 16)

Where:

θ_i - soil water content in the root zone (mm3 mm⁻ 3) on day i;

θ_{i-1} - soil water content in the root zone on day i-1 (m3. m⁻ 3);

P_i - rainfall on day i (mm);

Q_{ri} - surface runoff on day i (mm);

I_{ni} - (liquid) irrigation water on day i (mm), i.e. the amount of irrigation water that has actually infiltrated to be stored in the root zone;

ET_{ci} - crop evapotranspiration on day i (mm);

DP_i - deep percolation on day i (mm);

GW_i - accumulated capillary rise flow on day i (mm).

Z_{ri} - root length on day i (mm).

Based on the water content of the soil in the root zone (θ_i), the model calculates the net irrigation water (LL) required when the designated limit of restriction is reached.

In this work, this limit was defined as occurring when the water in the easily usable soil reserve (volume of soil considered, or square prism of soil) reaches a volume of 55% consumed by the plants (p=0.55). This volume must be replenished until the soil moisture (square prism of soil considered) reaches field capacity (FC).

Crop evapotranspiration (ETc) was calculated based on the product of reference evapotranspiration (ETo) and the corn crop coefficient (Kc) at each of its stages of development, according to Allen et al., (1998).

As already mentioned, reference evapotranspiration in the ISAREG model is calculated using the Penman Monteith equation, standardized by the FAO.

The average values for the crop coefficient (Kc) were obtained from FAO bulletin 56.

The contribution of groundwater (Gi) was not taken into account due to the lack of information on the depth of the water table, so it was considered that the water table was very deep and did not interfere in the water balance of the soils in question.

In calculating the water balance, the soil water available for plants is related to three storage levels (Figure 14): the excess water zone (θs), above field capacity, the optimum yield zone, between field capacity (θcc) and the optimum soil water limit (θp), where water is actually available for the crop and finally, the water deficit zone, between the optimum limit of water in the soil (θp) and the permanent wilting point (θpmp), where there is still water available for the plant, but it requires more energy to be extracted by the roots.

The size of the three zones of the reservoir in the soil varies from crop to crop, depending on the stage of development, root depth and the sensitivity of the crop to water deficit, expressed by the "p" fraction of depletion without stress, in the case of maize p=0.55.

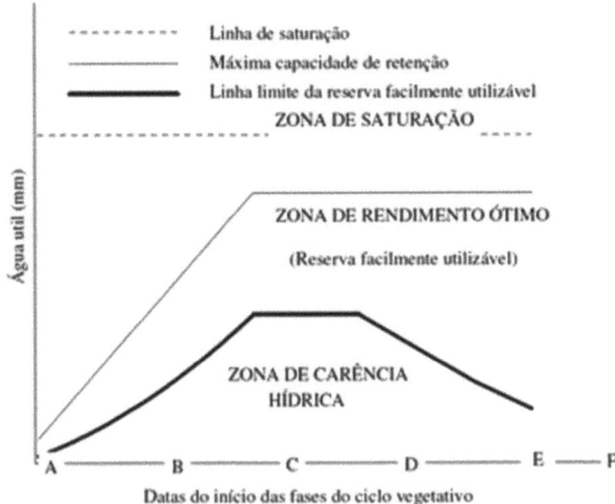

Figure 14 - Delimitation of the zones of maximum yield and water deficiency. Zone A, from planting to 1ª irrigation, zone B, start of vegetative growth, zone C, start of flowering, zone D, fruit formation, zone E, ripening phase and zone F, harvest.

The upper limit of the optimum yield zone (Rmàx, mm) corresponds to the amount of total water available in the soil (CAD) at the root depth of the crop considered (z). In the ISAREG model, the upper limit of the optimum yield zone is given by equation 17 to obtain the current soil moisture when calculating the water balance:

$$\mathbf{Rmáx = 1000 * Z * (\theta CC - \theta PMP)}$$

(Equation 17)

The lower limit of the optimum yield zone (Rmin, mm) is given by:

$$\mathbf{Rmin = (1 - p) * Rmáx}$$

(Equation 18)

Where "p" is the fraction of available water depletion in the soil that causes a water deficit for the crop. We used p=0.55 for maize, according to the FAO 56 table in the ISAREG program's explanatory manual.

This lower limit of the optimum yield zone defines the limit of the optimum soil water level (θp), i.e. the minimum soil water content before crop water deficit. The difference between Rmax and Rmin is the amount of soil water actually available or usable by the plants (CAD, mm), which is given in the model by the following

65

equation:

$$CAD = (\theta CC - \theta PMP) * z$$

(Equation 19)

The data required by the ISAREG model to calculate the soil water balance was stored in several files controlled by a main menu. These included meteorological data for the city of Sao Miguel do Oeste such as: reference evapotranspiration (ETo) and rainfall, in mm, at each time interval for calculating the water balance; maximum and minimum temperatures, in °C; relative humidity, in %; solar radiation, in MJ m^{-2} day^{-1} ; and wind speed, in m s-i for calculating crop evapotranspiration (ETc) using the EPAP56 subroutine.

As for the maize crop, the data required by the model are: determining the dates of its stages of development; the crop coefficient (Kc); the effective root depth (m); the "p" fraction of available water depletion in the soil and the response factor and yield.

The database for the characteristic soils of the Far West region of the state of Santa Catarina was organized with data on the depth of the soil layers (m), the total available water capacity in each layer (mm) or the gravimetric water content at CC and PMP (%); the soil density of each layer (g cm^{-3}) and the depth and texture of the evaporative topsoil, which are necessary for calculating the crop coefficient (Kc) for the initial period of crop development.

Tables 7 and 8 show some data relating to the maize crop, soil types and tillage systems that were organized in the form of files to feed the routines for calculating the maize crop's water requirements under the conditions established by the ISAREG program.

Table 7 - Corn crop data used to apply the ISAREG model.

Phenological stage of the crop	Date	Prof. of roots (z) in meters (m)	Depletion factor (P)	Crop coefficient (Kc)
Stage A - Planting and start of 1[a] irrigation	15/9	0,10	0,55	0,30
Stage B - Start of vegetative growth	24/9	0,10	0,55	0,80
Stage C - Beginning of	15/11	0,60	0,55	1,10

66

intermediate phase I - Flowering					
Stage D - Beginning of intermediate stage II - Fruit formation		5/12	0,60	0,55	1,10
Stage E - Beginning of the final phase - maturation		4/1	0,60	0,55	1,10
Stage F - Harvesting		24/1	0,60	0,55	0,40

Source: ISAREG

Table 8 - Volumetric soil moisture data for each layer analyzed and soil density for these layers used to apply the ISAREG model.

Soil	Management	Depth (m)	Field capacity (%)	Point of Permanent wilt (%)	Apparent density (g cm^{-3})
Cambisol	*CM	0,00 - 0,10	34,42	20,68	1,14
		0,10 - 0,20	36,25	25,47	1,20
		0,20 - 0,40	35,66	22,59	1,04
		0,40 - 0,60	38,49	25,27	1,06
	*PP	0,00 - 0,10	41,63	30,56	1,13
		0,10 - 0,20	38,18	27,49	1,13
		0,20 - 0,40	39,86	28,41	1,13
		0,40 - 0,60	33,71	19,71	1,06
	*PD	0,00 - 0,10	37,80	28,83	1,18
		0,10 - 0,20	31,73	21,81	1,14
		0,20 - 0,40	34,90	20,97	1,09
		0,40 - 0,60	34,93	21,95	1,08
Nitossolo	*CM	0,00 - 0,10	39,05	29,51	1,14
		0,10 - 0,20	40,41	32,68	1,21
		0,20 - 0,40	38,48	30,18	1,23
		0,40 - 0,60	39,03	28,44	1,05
	*PP	0,00 - 0,10	41,50	33,39	1,26
		0,10 - 0,20	42,42	33,64	1,23
		0,20 - 0,40	41,88	31,29	1,18
		0,40 - 0,60	41,65	30,67	1,20
	*PD	0,00 - 0,10	41,11	32,62	1,26
		0,10 - 0,20	41,11	33,36	1,29
		0,20 - 0,40	44,59	33,71	1,14
		0,40 - 0,60	43,87	32,42	1,06
Latosol	*CM	0,00 - 0,10	34,74	23,48	1,05
		0,10 - 0,20	42,59	34,57	1,17
		0,20 - 0,40	40,71	30,98	1,13
		0,40 - 0,60	48,84	38,15	0,99
	*PP	0,00 - 0,10	40,16	31,29	1,08
		0,10 - 0,20	42,57	34,27	1,16
		0,20 - 0,40	43,35	34,19	1,11
		0,40 - 0,60	43,66	33,26	1,07
	*PD	0,00 - 0,10	38,39	28,09	1,10
		0,10 - 0,20	37,67	30,52	1,22
		0,20 - 0,40	43,00	33,33	0,83
		0,40 - 0,60	51,49	39,94	0,94

*CM = Minimum Cultivation; *PP = Perennial Pasture; *PD = Direct Planting

4. Results and discussion

The Results and Discussion section of this paper is organized in three different ways. The first evaluates the values of the irrigation rates by fixing the type of soil (Cambissolo, Nitossolo and Latossolo) and varying the management systems (minimum cultivation, perennial pasture and no-till). The second evaluates the behavior of the irrigation rates to be applied by fixing the soil management system and varying the soil types. Finally, it compares the variations in irrigation blade values for the period 2014-2036 and 1988-2010.

It is important to emphasize that all the analyses of soil water requirements were carried out using the same edaphoclimatic data for both study periods. Therefore, the coefficients that infer or not infer water differences in soils and their management relate to the physical-hydric variations present in each soil and in each management system evaluated.

4.1 - Assessment of the water requirements of each type of soil in each management system

This evaluation analyzed the net blade values in each type of soil (Cambissolo, Nitossolo and Latossolo) when subjected to the minimum cultivation, perennial pasture and no-till management systems, i.e. by fixing the type of soil and varying the management system.

To this end, irrigation was considered to take place when the readily available water content in the soil in the root zone of the corn crop reached 45% of its total value, i.e. when the crop consumed 55% of the total volume of readily available water, the soil should be irrigated until its humidity returned to field capacity (FC).

For these analyses, climatological data from the period 19882010 was used, provided by the Santa Catarina Center for Environmental Information and Resources and Hydrometeorology (EPAGRI/CIRAM).

Figures 15, 16 and 17 show the annual liquid manure values to be applied to Cambissolo, Nitossolo and Latossolo, respectively, when subjected to minimum tillage, perennial pasture and no-till management systems. These evaluations were carried out using climatological data for the corn growing season in the western region of the state of Santa Catarina, which was defined as sowing date on September 15th and harvest date on January 24th of the following year.

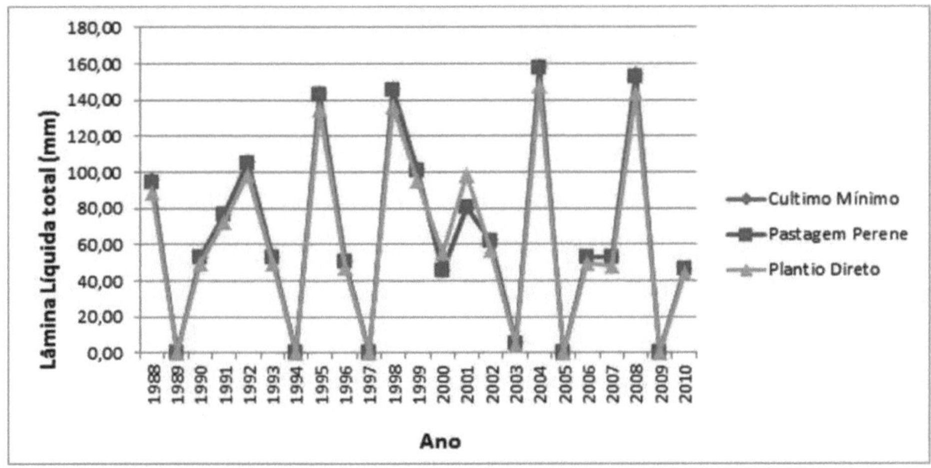

Figure 15 - Annual values (1988-2010) of irrigation water to be applied to the corn crop in a Cambissolo subjected to the minimum cultivation, perennial pasture and no-till management systems in the region of São Miguel do Oeste, Santa Catarina.

Table 9 - Descriptive statistics for estimates of total annual irrigation applied to a cambisol under three different managements in the period 1988-2010.

Total annual irrigation of a Cambissolo under three managements Management No. obs. Mean (mm) Standard deviation Variance Median

CM*	18	82,57	44,28	1960,9	69,75
PP*	18	81,96	44,03	1939,1	69,35
PD*	18	78,78	41,2	1704,7	64,7
CM*=	Cultivation Minimum,	**PD*=**	Direct Planting	and	Perennial Pasture
	ion		**PP*=**		

In Figure 15 above, through the analysis of total irrigation rates for the corn growing season in this region, estimated by the ISAREG model, it can be seen that the three management systems are practically the same over the period studied. Looking at Table 9, it is possible to see that, on average, the Direct Planting (PD) management system has slightly lower water requirements than the other management systems studied (CM and PP). Furthermore, the standard deviation in the table shows that the irrigation rates of the PD management system are the ones with the smallest variations in relation to the other managements. This small reduction in the water requirements of the cambisol when subjected to the no-till management system can be justified by the fact that this system, defined as the process of sowing in unturned soil and with the presence of mulch from the previous crop(s), in which the seed is placed in furrows or pits, prevents surface sealing due to the impact of raindrops. As a result, it hinders and reduces surface run-off, increasing the time that water remains on the soil and thus the infiltration of this water, as well as drastically reducing erosion. In addition, the mulch present on the surface also acts to protect the soil against the effect of the sun's rays, reducing evaporation, soil temperature, temperature range and wind action and thus increasing water availability to plants, reducing the need to replace this water in the soil (EMBRAPA, 2013).

It should be noted that the soil classified as Cambissolo is a shallow, undeveloped soil that still shows characteristics of the original material (rock), which decreases the permeability and exploration depth of the crop roots in this soil. As a result, the roots of the plants have less soil volume to explore, thus reducing the availability of water and nutrients, as well as affecting their ability to support themselves.

According to Hakansson et al. (1998), the lower development of the root system results in a smaller volume of soil explored by the roots and, consequently, lower absorption of water and nutrients.

Figure 15 also shows that the years in which there was no need to irrigate the soil, i.e. zero irrigation, coincide in all management systems, indicating that in these years the rainfall in the region was sufficient to meet the water needs of the maize crop in the region, assuming that the time for irrigation was defined when the plant had consumed 55% of the easily usable water reserve in the soil, for all management systems.

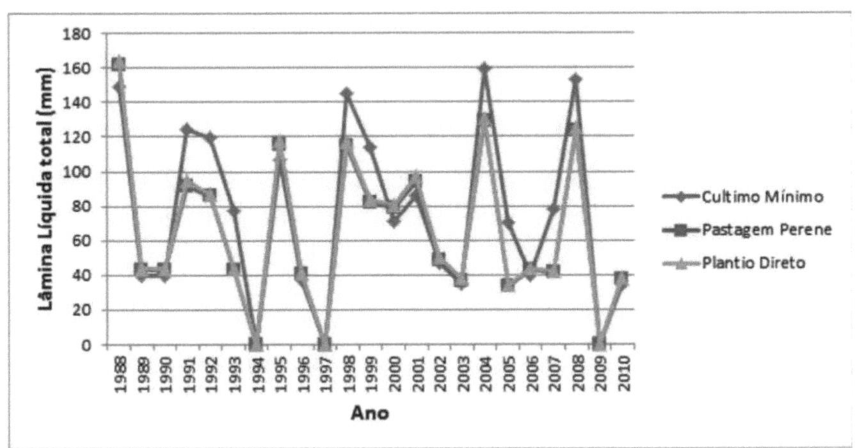

Figure 16 - Annual values (period 1988-2010) of irrigation water to be applied to the corn crop in a Nitossolo subjected to minimum cultivation, perennial pasture and no-till management systems in the region of Sao Miguel do Oeste, Santa Catarina.

Table 10- Descriptive statistics of the estimates of total annual irrigation rates applied to a nitossolo under three different managements in the period 1988-2010.

Total annual irrigation of a Nitossolo under three managements

Management	Obs. no.	Average (mm)	Standard deviation	Variance	Median
CM*	20	86.285	44.345	1966.5	77.35
PP*	20	74.395	38.774	1503.4	63.85
PD*	20	75.825	39.393	1551.8	65.4

CM*= Minimal Cultivation, PD*= Direct Planting and PP*= Perennial Pasture

Figure 16 and Table 10 show that the annual net irrigation rates to be applied to the Nitossolo were higher in the minimum cultivation system when compared to the other systems studied here. It is also possible to see from the standard deviation that CM is the soil with the greatest variations in water levels.

Comparing the suggested blade values for the PD and PP management systems, it can be seen that in both systems the blade values to be applied over time are practically the same in this type of soil. Furthermore, according to Lanças (1988), even though cultivation systems provide favourable conditions for the growth and production of crop plants, intensive tillage and the movement of vehicles, animals and machinery, which are generally heavy, over the soil, contribute to the formation of compacted layers.

Silveira (1988) states that systems which aim for a minimum of tillage and which provide a maximum of plant residues on the surface, guarantee better infiltration and water retention, better soil structure, porosity, content and distribution of organic matter.

In Figure 16, it can be seen that the maize crop grown on Nitosolo under the CM tillage system required a greater application of irrigation during the period under study, due to the compaction that occurs in the surface layers of this soil when subjected to such management, as a result of the mechanization system used in this tillage. This is because the CM tillage system requires at least one mechanization process - scarification, heavy harrowing or tillage with a rotary hoe - when compared to the PD process, which only uses a seeder to mechanize the soil, which carries out this process directly in the furrow where seeds and fertilizers are deposited.

In scarification, which takes place in CM, the soil is prepared without inversion and maintains an average of 70% vegetation cover. Heavy harrowing consists of preparing the soil using disk harrows. In this variation of minimum tillage, the soil is inverted and the vegetation is chopped up and incorporated into it.

The minimum tillage system with a rotary hoe consists of cutting the soil into small fractions using rotating blades, causing high soil mobilization, implying destruction of the structure and pulverization of the soil. The soil surface is left with little or no vegetation, favoring the formation of a surface crust.

In view of this, it is possible to see that the mechanization processes, even if reduced in the CM system, are still superior to the mechanization processes in the PD, which causes greater compaction of the topsoil when compared to the PD and PP, making it difficult for water to infiltrate the soil. This was confirmed by Meyer and Mannering (1961).

According to Melo et. al. (2007), turning over the soil reduces soil density and increases soil porosity. Hakansson et al., (1998), states that more compacted soils imply less development of the root system and consequently less absorption of water and nutrients by plants. In addition, soils with compacted surface layers lead to greater water losses through surface runoff.

Another fact that may explain the greater water demand of the nitossolo under CM is related to the amount of mulch present on the surface of the soil. When the CM process is carried out with a rotary hoe, or even harrowing, the straw present in the soil from the previous crop is incorporated into the soil, leaving it partially bare, which favors water losses in the soil through evaporation and also favors soil erosion. In addition, the mulch on top of the soil makes it difficult for water to percolate into the soil, allowing it to remain there for longer and thus favoring infiltration. Mulch on the surface also intensifies microbial activity, which is largely responsible for decompaction and total soil porosity.

Mclsaac et al. (1987), evaluating the losses that occurred in the soil through the crop rotation system between maize and soya, found greater losses through surface runoff during maize cultivation, when the straw from soya in the previous period was less on the soil.

Figure 17 below shows the variations in the total water requirement of the maize crop when grown on a latosol on the different soil management systems (CM, PP and PD). In both figure 17 and table 11, it is possible to see small variations in the water requirements of this soil when subjected to PP management compared to CM and PD management, where in almost all the years of the study, it was possible to see reductions in the water requirements of maize in a latosol under the PP system. In this management system, pasture is grown during the winter and grazed by animals in a Voisin system. Voisin Rational Grazing (VRP) is a pasture management system that involves dividing the pasture area into several plots, where they are made available for cattle grazing individually and sequentially, with the aim of increasing pasture productivity, greater root development, which avoids soil erosion and compaction due to animal trampling. Next, the pasture is desiccated and then, in the same area, in the summer, the corn crop is sown in the soil, under direct planting of the seed.

Given the management system described above, it is possible to infer that the soil's water needs are reduced due to some of the advantages that this management gives the soil in terms of its infiltration capacity and water storage, where the frequent deposition of organic material by the cattle increases soil fertility and microbial activity. In addition, the presence of straw and roots from the pasture also increases the soil's organic matter content, which leads to an increase in total porosity and better soil structure, favoring water percolation and allowing water to remain in the soil for longer, thus increasing its infiltration capacity. According to Embrapa (2005), the decomposition of the roots also creates a network of channels in the soil, which is very important for gas exchange, and a downward movement of water, allowing water to percolate into deeper layers of the soil.

With regard to the slightly higher water requirements that the latosol showed under CM management compared to PD and PP, it is possible to infer that this was due to the higher degree of soil compaction.

In some years, it is possible to see an increase in the need for soil irrigation on CM when compared to the other management systems under study, highlighting the assertions that mechanization prior to sowing generates superficial soil compaction, which makes it difficult for water to infiltrate the soil and, as a result, requires a greater volume of irrigation during the maize crop cycle.

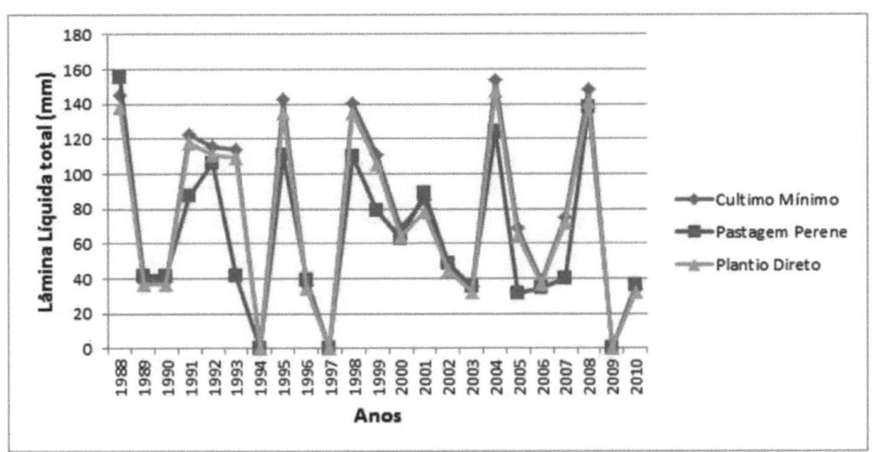

Figure 17 - Annual values (1988-2010) of irrigation water to be applied to the corn crop in a Latosol subjected to minimum cultivation, perennial pasture and no-till management systems in the region of Sao Miguel do Oeste, Santa Catarina.

Table 11- Descriptive statistics for estimates of total annual irrigation applied to a latosol under three different managements in the period 1988-2010.

Total annual irrigation of a Latosol under three managements Management No. obs. Mean (mm) Standard deviation Variance Median

	No. obs	Mean (mm)	Standard deviation	Variance	Median
CM*	20	87,93	45,03	2028,0	80,25
PP*	20	72,62	39,74	1579,4	55,35
PD*	20	83,66	43,19	1866,1	74,9

CM*= Minimal Cultivation, PD*= Direct Planting and PP*= Perennial Pasture

Statistically comparing the mean values of total irrigation required for annual maize cultivation in the following soils: cambisol, nitosol and latosol when subjected to three different managements: minimum cultivation, no-till and perennial pasture, it can be seen that they do not differ significantly by the test of

Kruskal-Wallis at the 5% probability level, as shown in Table 12.

Table 12- Kruskall-Wallis test for total annual irrigation of the corn crop.

Total annual irrigation for the corn crop (mm)

CAMBISOL			NITOSSOLO			LATOSSOLO		
System	No. obs	Median	System	N° obs	Median	System	No. obs	Median
			CM	2086	.28			
CM	18	82.57 ns	ns			CM	20	87.93 ns
PP	18	81.96 ns		PP	2074	.39	PP	20	72.62 ns

		ns				
PD	18	78.78 ns		PD2075	.82PD	20	83.66 ns
		ns				

3.3 - Evaluation of the water requirements of different soils when subjected to the same management system

Soil texture refers to the relative proportion in which the different particle sizes are found in a given mass of soil. Specifically, it refers to the relative proportions of sand, silt and clay particles or fractions. This physical property of the soil is the one that changes the least over time. Soil texture is a very important factor in irrigation because it has a direct influence on the rate of water infiltration into the soil, on aeration, on water retention capacity, on nutrients, as well as on the adherence or cohesive force of soil particles (EMBRAPA, 2003). The levels of sand, silt and clay in the soil have a direct influence on soil preparation and planting, making it easier or more difficult for machines to work. It also influences the choice of irrigation method to be used.

Sandy textured soils are considered light soils and are classified as those with a sand content of more than 70% and a clay content of less than 15%; they are permeable, light, have a low water retention capacity and a low organic matter content. They are highly susceptible to erosion, requiring special care in the replenishment of organic matter, soil preparation and conservation practices. As a characteristic, they have a high capacity for infiltrating water into the soil, but a low capacity for retaining it. As a result, there are high percolation losses.

Medium-textured soils are soils with a certain balance between sand, silt and clay content. They usually have good drainage, good water retention capacity and a medium erodibility index.

Clay-textured soils are considered to be heavy soils with clay contents of more than 35%. They are characterized by their low permeability and high water retention capacity.

Although they are more resistant to erosion, they are highly susceptible to compaction, which means that special care must be taken when preparing them, especially with regard to the moisture content, where the soil must have a crumbly consistency. In view of this, table 7, which shows the textural data of the soils evaluated in this study, makes it possible to classify nitossolo and latossolo as clay-textured soils, as they have around 50% of their total textural percentage made up of clay.

The cambisol, on the other hand, can be classified as medium-textured, where the sand fraction is less than 70% and the clay fraction less than 35%.

Evaluating the answers obtained regarding the total annual irrigation that should be provided to the three soils studied (cambisol, latosol and nitosol), when subjected to the management known as minimum cultivation (Figure 18), it is possible to see that the cambisol shows slightly lower values of water requirements over the years studied compared to the nitosol and latosol. This is probably because they are medium-textured soils with good drainage and water retention capacity (EMBRAPA, 2003). In addition, they are shallow to deep soils, which characterize soils that can both soak easily and dry out quickly, and are very susceptible to local climatic variations.

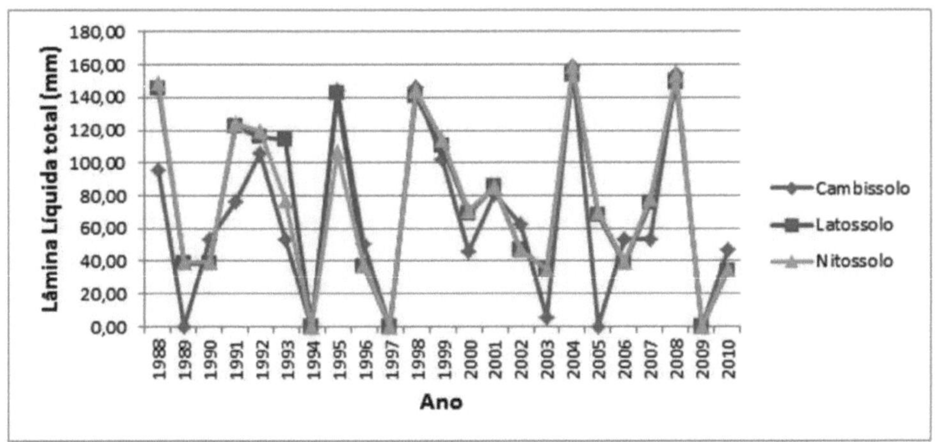

Figure 18 - Variations in the water requirement of the corn crop in three different soils (cambisol, yatossol and nitossol), when subjected to minimum cultivation.

Table 13- Descriptive statistics of the estimates of total annual irrigation applied to corn in three different soils, when subjected to minimum cultivation in the period 1988-2010.

Total annual irrigation of three soils subjected to Minimal Cultivation management

Soil	Obs. no.	Average (mm)	Standard deviation	Variance	Median
CAM*	18	82,57	44,28	1960,9	69,75
LAT*	20	87,93	45,03	2028	80,25
NIT*	20	86,28	44,34	1966,5	77,35

CAM= cambisol, LAT= latosol and NIT= nitosol

On the other hand, latosol and nitosol, which have a heavy texture with a high percentage of clay in their structure, showed slightly higher annual water requirements when subjected to minimum tillage, and this is due not only to physical factors, where they are soils highly susceptible to compaction, considered a very important factor for the percolation and infiltration of water into the soil, They are also soils with a good water retention capacity, but because the clays have a high cohesive force between the soil particles and the water, the water is not fully available to the plants, which makes it difficult for the crop to take advantage of the moisture in the soil. Clay soils (and those with a high organic matter content) retain water more strongly than sandy soils. This means more water in the soil, but not available (EMBRAPA, 2003).

In view of this, the CM infers a greater need for water in the nitossolo and latossolo when compared to the cambissolo.

Figure 19 shows the variations in annual water requirements that occur in the three soils when subjected to perennial pasture management under corn cultivation during the winter period. From spring to summer, these soils are cultivated with corn under conventional management.

In this, it is possible to see greater variation in annual water requirements in the cambisol, while the nitosol and latosol show a similar pattern over the years studied (1988-2010).

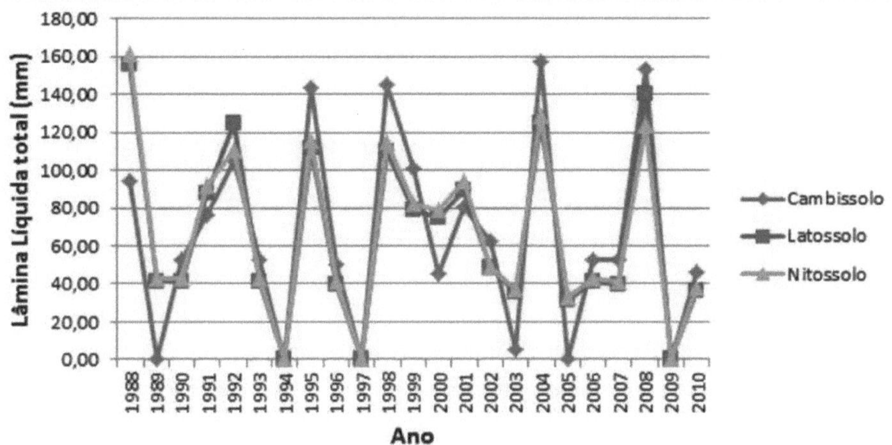

Figure 19 - Variations in the water requirement of the corn crop in three different soils (cambisol, latosol and nitosol), when subjected to Perennial Pasture management.

Table 14- Descriptive statistics for estimates of total annual irrigation applied to corn in three different soils, when subjected to Perennial Pasture management in the period 1988-2010.

Total annual irrigation of three soils under Perennial Pasture management

Soil	Obs. no.	Average (mm)	Standard deviation	Variance	Median
CAM*	18	81,96	44,03	1939,1	69,35
LAT*	20	75,63	42,45	1801,9	61,6
NIT*	20	74,39	38,77	1503,4	63,85

CAM= cambisol, LAT= latosol and NIT= nitosol

These results can be explained by the clay texture of the nitossolo and latossolo, as opposed to the medium texture of the cambissolo. As already mentioned, the clayier soils have a high capacity for storing water, which is retained between the particles by cohesive forces. Cambisols, considered to be shallow to deep soils, have good drainage, but due to the small space available for plants, these soils can either waterlog easily or dry out quickly. Therefore, climatic factors such as high temperatures can have a significant influence on the water requirements of these soils when subjected to such management, as they can accelerate or favor soil evaporation.

In Figure 20, even with some variations, it is still possible to see similar trends in the water requirements of the different soils when subjected to no-till management.

Just a few points from the historical series of analyses should be highlighted: in figure 20 below, it is possible to see that the latosol shows a slight tendency towards lower annual water requirements compared to the other soils in the study and this is probably due to its clayey to very clayey texture, which allows this soil to have a greater water retention capacity and even less water loss through evaporation into the environment, something that is further favored by the mulch present on the soil through no-till farming (PD).

The nitossolo, which is also considered to be a clay-textured soil, has slightly higher annual water requirements when compared to the latossolo, and this effect can be explained by the fact that it is a good B textural soil, and a little shallower than the latossolo, thus allowing it to have higher evaporation losses to the environment than the Iatossolo.

In relation to the cambisol, it can be seen that this soil has more years in which there is no need to add water to the soil than the others and also has the highest irrigation rates in other years of the series, and this is due to it being classified as shallow soil, which is very susceptible to bad weather, i.e. very little liquid rainfall is able to bring the soil up to the hydric condition defined in this study, which is to irrigate the soil when 55% of the AFD has been consumed by the plants and water it until it reaches CC again. In addition, temperature variations over the course of the series can also have a significant impact on the soil's water requirements. In addition, this soil has a higher sandy fraction than the others, which can lead to greater losses through deep drainage or even soil evaporation.

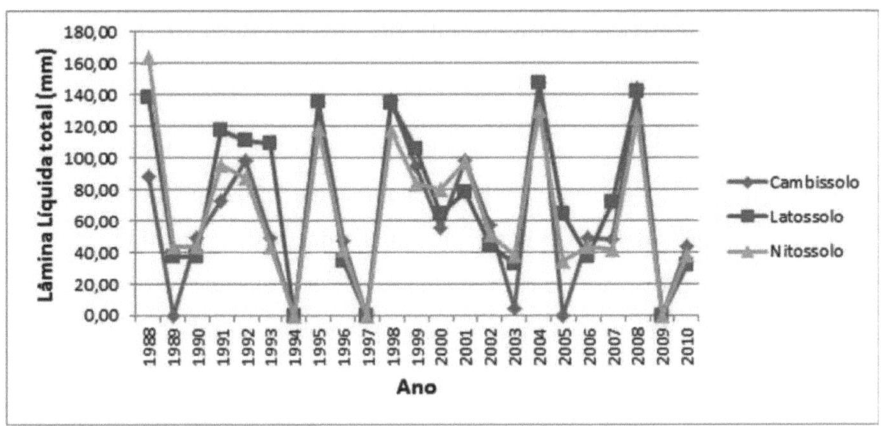

Figure 20 - Variations in the water requirement of the corn crop in three different soils (cambisol, latosol and nitosol), when subjected to no-till management.

Table 15 - Descriptive statistics for estimates of total annual irrigation rates applied to the corn crop in three different soils, when subjected to no-till management during the period 19882010.________________

Total annual irrigation of three soils under no-till management

Soil	Obs. no.	Average (mm)	Standard deviation	Variance	Median
CAM*	18	81,86	41,63	1733,1	80,45
LAT*	20	83,36	42,92	1842	74,9
NIT*	20	75,82	39,39	1551,8	65,4

CAM= cambisol, LAT= latosol and NIT= nitosol

Statistically comparing the mean values of total irrigation required for growing corn in different soils when subjected to the same soil management (CM, PP and PD), it can be seen that they do not differ significantly by the Kruskal-Wallis test at 5% probability, as shown in Table 10.

Table 16- Kruskall-Wallis test for total annual irrigation of the corn crop.

Total annual irrigation for the corn crop (mm)

Minimum Cultivation (CM)   Perennial Pasture (PP)   Direct Planting (PD)

Soil	No. obs	Median	Soil	No. obs	Median	Soil	No. obs	Median
Cambisol	18	69.75 ns	Cambisol	18	69.35 ns	Cambisol	18	64.70 ns
Latosol	20	80.25 ns	Latosol	20	61.60 ns	Latosol	20	74.90 ns

3.4 - Assessment of future water needs for the far west of the state of Santa Catarina

Crop water consumption is known to be directly affected by climate change. According to the definition given by the Intergovernmental Panel on Climate Change (IPCC), climate change is defined as significant alterations in the state of the climate, which can be identified by variations in averages and/or measures of dispersion of climate variables, and which persist for long periods, typically decades or even longer (ABREU and PEREIRA, 2007).

In the 100 years between 1905 and 2006, the Earth's air temperature rose by 0.74 °C, while in the 50 years between 1956 and 2005 it rose by 1.30 °C, almost double the previous figure (IPCC, 2007).

Due to this report and estimates of temperature increase, this work also decided to assess whether there would be an increase in the water requirements of the corn crop in the far west of the state of Santa Catarina for the period 2014 to 2036 when compared to the water requirements obtained for the period 1988 to 2010.

In order to carry out this evaluation, all the years in which there was no need to irrigate the corn crop were eliminated, i.e. where the total annual value of LL to be made available to the crop was zero, both in the 1988-2010 series and in the 2014-2036 series. This measure was taken to eliminate distortions that this value could infer in relation to the overall average.

Next, an average was made of the total annual LL values available for each annual series and, using these, comparisons were made as to the water requirements of the series. All the total annual WL and average values are in the appendix of this study.

It is important to point out that this analysis was not statistically verified due to the fact that there is no correlation between the series, i.e. the irrigations that occur in one year do not affect the water needs of the following year.

Through the evaluations obtained in this work and presented in Table 11 below, it is possible to verify that these results are contrary to the IPCC's (2007) global temperature increase assumptions for the far west region of Santa Catarina, because evaluating the average values obtained, in general, the vast majority of managements applied to the soils characteristic of this region indicate a reduction in the volume of water to be applied to the soil to meet the needs of the corn crop. Only PD when applied to cambisol and PP when applied to latosol and nitosol indicate an increase in water requirements for the corn crop. While the other combinations of soil and management indicate a decrease in this variable, CM applied to latosol can imply a reduction of up to 9.5 mm of water applied on average per year, which corresponds to a difference of approximately 11%. A similar reduction also occurs with CM applied to nitossolo, where the difference is around 8.8%.

Table 17- Evaluation of total annual LL averages (mm) between the years 1988-2010 and 2014-2036

Soil	Manage ment	Series			
		1988-2010	2014-2036	Difference (mm)	Difference (%)
Cambisol	CM	82.6	82.4	-0,2	-0,2
	PP	82	82	0,00	0,0
	PD	81.9	82.8	0,9	1,1
	average	82,17	82.4	0,23	0,28
Latosol	CM	87.9	78.4	-9,5	-10,8
	PP	75.6	78.5	2,9	3,8
	PD	83.4	78.8	-4,6	-5,5
	average	82,3	78,6	-3,7	-4,5
Nitossolo	CM	86.3	78.7	-7,6	-8,8
	PP	74.4	79.6	5,2	6,9
	PD	75.8	75.8	0,0	0,0
	average	78,8	78,0	-0,8	-1,01

Evaluating the table by management, it can be seen that CM will result in water reductions in all the soils in the study. PP, with the exception of cambisol, will result in an increase in total annual WL of up to 6.9%. On the other hand, PD in the cambisol will mean an increase in annual LL of 1.1%, while in the nitossolo there will be no need for changes and in the latossolo there will be a reduction in the necessary water resources.

In general, it is possible to indicate that for the far western region of Santa Catarina there will be a reduction in the water needs of the corn crop for the 2014-2036 period of approximately 5.5% when compared to the water needs of the 1988-2010 period.

These observations can be justified by information provided by the Intergovernmental Panel on Climate Change (IPCC) itself, which estimates that the primary warming will occur in the oceans, indicating that the temperature of the ground should have an insignificant amplitude, less than 0.006 °C per decade. Another fact that may justify these observations refers to future assumptions of an increase in precipitation and its frequency for South America, mainly in cities with altitudes above 700 m, which is evident in the results obtained in this work, since the altitude of the city under study, Sao Miguel do Oeste, is 946 m (IPCC, 2007).

5. Conclusions

According to the results obtained, it was possible to conclude that:

When evaluating the different water needs of the maize crop when grown in a cambisol subjected to three different managements (CM, PP and PD), it was found that the water needs of this soil are very similar, regardless of the management applied to it.

When evaluating the water requirements of maize when grown on nitossolo soil and subjected to three different management methods: Minimum Cultivation (MC), Direct Planting (PD) and Perennial Pasture (PP), it is possible to see similar water requirements for maize when grown on nitossolo under PP and PD management. CM, on the other hand, implies a greater need for water than the other crops shown here.

By checking how soil management affects the water requirements of the maize crop in the different soils studied (cambisol, latosol and nitosol), it can be seen that the management of perennial pastures results in similar water requirements to maize when grown in latosol and nitosol soils.

When comparing the water requirements of the maize crop when grown in different soils (cambisol, latosol and nitosol) and subjected to the same management, it can be seen that the variation in soils does not lead to significant differences in the water requirements of the maize crop.

When comparing the water requirements of maize when grown on the same soil but under different management (CM, PP and PD), it can be seen that soil management does not lead to significant differences in the water requirements of this crop.

After applying the model of this study, it was found that in the far west of Santa Catarina there will be a reduction in the water needs of the corn crop for the period 2014-2036 compared to the water needs of this crop in the period 1988-2010.

In the 2014-2036 period, the Yatosol shows itself to be the soil that will need the least LL of water for growing corn when compared to the LL to be needed for the crop grown on the other soils present in the study. The cambisol, on the other hand, shows practically no variation in future water requirements (2014-2036) for maize compared to the water requirements in the 1988-2010 period.

6. BIBLIOGRAPHICAL REFERENCES

AGUIAR, Raquel. Drought: how to live with this phenomenon? Rural Extension and Sustainable Development. Porto Alegre, v.1 n.4, p. 11-14, 2005.

AGRIANUAL: yearbook of Brazilian agriculture. Sao Paulo: FNP, 2009. p.405- 410.

NATIONAL WATER AGENCY (ANA). Caderno de Recursos Hidricos, disponibilidade e demandas de recursos hidricos no Brasil. Brasilia-DF. 2005.

USP NEWS AGENCY. Crop-livestock integration is a sustainable alternative. Available at: < http://www.usp.br/agen/?p=141604> Accessed on: June 18, 2013.

ALLEN, R. G. Penman for all seasons. Journal of Irrigation and Drainage Engineering, New York, v. 112, n. 4, p. 348-368, 1986.

ALLEN, R. G.; PEREIRA, L. S.; RAES, D.; SMITH, M. Crop evapotranspiration: guidelines for computing crop water requirements. Rome: FAO, 300 p. 1998 (FAO. Irrigation and Drainage Paper, 56).

ALMEIDA, F. S. Influence of no-till mulch on soil biology. In: FANCELLI, A. L. (Coord.) Atualizaçao em plantio direto. Campinas: Cargill Foundation, p.103-147. 1988.

ALVARENGA, R. C., NOCE, M. A. Crop and livestock integration. Embrapa, National Maize and Sorghum Research Center. Doc. 7. Sete Lagoas, MG. 2005.

AMORIN NETO, M. da S. - Water balance according to Thornthwaite and Mather (1955), Technical Communication. Petrolina: EMBRAPA-CPATSA, 1989.

ANDRADE, F. H. Radiation and temperature determine maximum maize yields. Balcarce: National Institute of Agricultural Technology, 1992. 34 p. (Technical Bulletin, 106).

ARAÙJO, E.F.; GALVÂO, J.C.C.; MIRANDA, G.V.; ARAÙJO, R.F. Physiological quality of sweet corn seeds submitted to threshing, with different degrees of humidity. Revista Brasileira de Milho e Sorgo, v. 1, n.2, p. 77-86, 2002.

ARNON, I. Mineral nutrition of maize. Bern: International Potash Institute, 1975.

452 p.

BERGER, J. Maize Production and the manuring of maizes. s.l.: Center d'Estudo de l'azote, 1962. p. 38 - 41.

BUENO. Climate studies: Evapotranspiration. Instituto Federal Goiano, Campus Iporà, Goiàs, 2012.

CAMARGO, A. P. Contribution to the determination of potential evapotranspiration in the State of Sao Paulo. Bragantia, 21: 163-213, 1962

CAMEIRA, M.R.; FERNANDO, R.M.; AHUJA, L.; PEREIRA, L.S. Simulating the fate of water in field soil-crop environment. Journal of Hydrology 315: 1-24. 2005.

COITINHO, J.B.L. Mineral waters of Santa Catarina, 2000. Dissertation (Master's Degree in Civil Engineering) - Technology Center, Federal University of Santa Catarina, Florianópolis, 1998.

CONAB (National Supply Company) - Monitoring the Brazilian Grain Harvest - 2011/2012 Harvest. May 2012.

CHEN, F., JANJIC, Z. I., MITCHELL K. Impact of atmospheric surface layer parameterization in the new land-surface scheme of the NCEP mesoscale Eta model. Bound-Layer Meteor 85:391-421. 1997.

CHOU, S. C, BUSTAMNTE, J. F.; GOMES J. L. Evaluation of Eta model seasonal precipitation forecasts over South America. Nonlinear Process Geophys 12(4):537-555. 2005.

CHOU, S. C.; MARENGO, J. A.; LYRA, A. A.; SUEIRO, G.; PESQUERO, J. F.; ALVES, L. M.; KAY, G.; BETTS, R.; CHAGAS, D. J.; GOMES, J. L.; BUSTAMNTE, J. F.; TAVARES, P. Downscaling of South America present climate driven by 4-member HadCM3 runs. Climate Dynamics, v. 38, p. 635-653, 2011.

DALMAGO, Genei Antonio et al., The "drought" phenomenon from an agronomic perspective. 2009. Available at: <http://www.cnpt.embrapa.br/pesquisa/solos/>.

DAKER, A. Agua na agricultura: manual de hidràulica agricola; Irrigaçao e

Drenagem. 3ª ed. Rio de Janeiro: Freitas Bastos, 1970. 453p.

DOORENBOS, J.; PRUITT, W.O. Guidelines for predicting crop water requirements. Irrigation and Drainage Paper, n. 24, 2ed. FAO. Rome. 1977. 156p.

DOORENBOS, J.; KASSAM, A.H. Efectos del agua sobre el rendimiento de los cultivos. Rome: FAO. FAO Study. Water and Drainage, 33. 1979. 212p.

DOORENBOS, J.; PRUITT, WO Crop Water Requirements. Rome: FAO, 1980. 194p. Water and Drainage, n.24.

DOORENBOS, J.; PRUITT, W.O. Guidelines for Predicting Crop Water Requirements. Irrigation and Drainage 24 (fourth ed.), FAO, Rome. 1992.

DOORENBOS, J.; KASSAM, A. H. Effect of water on crop yields. FAO Studies - Irrigation and Drainage n. 33, 1994. 306p. (translated by Ghevi, H. R. et al., - UFPB)

DUARTE, J. de O. Economic Importance. In: CRUZ, J. C.; VERSIANI, R. P.; FERREIRA, M. T. R. (Eds.) Cultivo do milho. Sete Lagoas: EMBRAPA - National Maize and Sorghum Research Center. 2000.

EMBRAPA. National Soil Research Center (Rio de Janeiro, RJ). Manual of Soil Analysis Methods. 2ed. Rio de Janeiro, 1997. 212p.

EMBRAPA. Irrigated cotton cultivation. Available at:

<http://sistemasdeproducao.cnptia.embrapa.br/FontesHTML/Algodao/AlgodaoIrriga d o/solos.htm> Accessed: Jun. 2013.

EMBRAPA. Reconnaissance survey of the soils of the state of Santa Catarina. Rio de Janeiro, 2004.

EMBRAPA. Crop-Livestock Integration. Sete Lagoas, MG. 2005.

EMBRAPA. Soils of Brazil. Embrapa Soils, 2013.

EMBRAPA. Embrapa Information and Technology Agency. Available at:

<http://www.agencia.cnptia.embrapa.br/gestor/cana-de-acucar/arvore/CONTAG01_85_22 122006154841.html> Accessed June 17, 2013.

EMBRAPA CORN AND SORGHUM. Available at: <http://sistemasdeproducao.cnptia.embrapa. br/FontesHTML/Milho/CultivodoMilho_2ed/mandireto.htm> Accessed: Jun. 2013.

EPAGRI CIRAM - Information Center for Environmental Resources and Hydrometeorology of Santa Catarina. Available at: <http://ciram.epagri.sc.gov.br> Accessed on: 28 Sept. 2011.

EPAGRI. Agricultural Research and Rural Extension Company of Santa Catarina. Santa Catarina Agricultural Survey 2002-2003. Available at: <http://cepa.epagri.sc.gov.br/Dados_do_LAC/lac_indice.htm>. Accessed on: July 15, 2013.

Ek MB, Mitchell KE, Lin Y, Rogers E, Grummen P, Koren V, Gayno G, Tarpley JD (2003) Implementation of NOAH land surface advances in the National centers for environmental prediction operational mesoscale Eta model. J Geophys Res 108:8851. doi: 10.1029/2002JD003246

FAO - Food and agriculture organization of the United Nations - Available at: http://www.fao.org>

FANCELLI, A.L. Influence of defoliation on plant and seed performance of maize (Zea mays L.). Piracicaba: ESALQ-USP, 1988. 172p

FANCELLI, A. L.; NETO, D. D. Corn production. Guaiba/RS Brazil, 2000.

FANCELLI, A. L. Update on no-till farming. Campinas: Cargill Foundation, 1985. 343p.

FANCELLI, A. L. No-till farming. Piracicaba: USP, ESALQ, Dept. of Agriculture, 1987. 89p.

FEHIDRO - State Water Resources Fund. State Water Resources Plan for Santa Catarina - PERH/SC. Vol. Iv - RH 1 - Far West - Volume I. Santa Catarina/ Brazil, 2008.

FERRAZ, E. C. Corn physiology. In: Brazilian Potash Institute. Culture and fertilization of maize. Sao Paulo, 1966. p. 369- 379.

FIESC. Federation of Industries of Santa Catarina. Web Guide. 2006.

FNP CONSULTORIA and AGROINFORMATIVOS, AGRIANUAL 2002. Anuàrio da agricultura brasileira, Sao Paulo, 2002, p. 417-437 (Maize).

FRATTINI, J. A. Maize cultivation: summary instructions. Campinas: CATI/ COT, 1975. 26p.

GODOY, L.J.G. Nitrogen management in corn (Zea mays L.) in sandy soil based on the relative chlorophyll index. 2002. 94p. Dissertation (Master's Degree in Agriculture) - Faculty of Agronomic Sciences, São Paulo State University, Botucatu, 2002.

GEO BRASIL RECURSOS HiDRICOS - National Water Agency, Brasilia/ DF. Available at: <http://www2.ana.gov.br> Accessed on: September 24, 2011.

GOMIDE, R. L.; OLIVEIRA, C. S. G.; FACCIOLI, G. G. Prototype of an automatic weighing lysimeter for greenhouse studies. In: Congresso Brasileiro de Agrometeorologia, 10, Piracicaba, 1996. P 225-227.

GUADAGNIN C. A.; TIMM L. C.; TAVARES V. E. Q.; LOUZADA J. A. S.; Validation of the ISAREG Model for Simulating the Water Balance in Cambisols in the Far West of Santa Catarina. XIV ENPOS. Pelotas/RS. 2012.

GUIMARAES, J. W. A. Management of irrigated cotton in the Jaguaribe-Apodi project, using a computer model. Master's dissertation. Fortaleza, 1993, 182p.

HANWAY J.J. How a corn plant delelops. Ames-Iowa: Iowa State University, 1966 (Special Report, 48).

HAKANSSON, I.;STENBERG, M.; RYDREBERG, T. Long-term experiments with different depths of mouldboard plowing in Sweden. Soil and Tillage Research. Amsterdam, v.46, p.209-223, 1998.

HARGREAVES, G. H.; SAMANI, Z. A. Reference crop evapotranspiration from temperature. Journal of Applied Engineering in Agriculture, St Joseph, v.1, n.2, p.9699, 1985.

IBGE. Brazilian Institute of Geography and Statistics. 2000 Demographic Census. Available at: <http://www.ibge.gov.br>. Accessed on: 07 May 2013.

IBGE - Brazilian Institute of Geography and Statistics. Plant Production. Accessed on Nov. 21, 2011. Available at:

<http://www.sidra.ibge.gov.br/bda/tabela/listabl.asp?c=283&z=t&o=3>

INTERGOVERNMENTAL PANEL ON CLIMATE CHANGE. Climate change 2007: the physical science basis. Summary for policymakers. IPCC: Geneva, 2007. Available at: <http://www.ipcc.ch>. Accessed on: Feb. 2013.

IPCC, Climate Change 2007: Synthesis Report - Summary for Policymakers. An Assessment of the Intergovernmental Panel on Climate Change. 2007.

JANJIC Z. I. The step-mountain Eta coordinate model: further developments of the convection, Viscous sub layer and turbulence closure schemes. Mon Wea Rev 122: 927-945. 1994.

JENSEN, M. E. Water consumption by agricultural plants. In: Kozlowski, T.T., Water deficit and plant growth, vol. 2, Academic Press, New York, 1968.

JOBIM, C. I., LOUZADA, J. A. Performance evaluation of the ISAREG water balance simulation model. Pesquisa Agropecuâria Gaûcha, Porto Alegre, V. 15. N.2, P.91-98. 2009.

KINIRY, J. R.; BONHOMME. Predicting maize phenology. In: HODGES, C. (Ed.) Predicting crop phenology. Boca Raton: CRC Press, 1991. p.115-131.

KOPPEN, W.; GEIGER, R. Klimate der Erde. Gotha: Verlag Justus Perthes. 1928.

LIBARDI, P. L. Soil Water Dynamics. Piracicaba, 2004.

LIU, Yujie et al., Model validation and crop coefficients for irrigation scheduling in the North China Plain. Agric. Water Manag. v.36. 1998. p.233-246.

MARCELINO, E. V.; GOERL, R. F.; RUDORFF, F. M. Space-time distribution of flash floods in Santa Catarina (period 1980 - 2003). In: Brazilian Symposium on

Natural Disasters, 2004, Florianópolis. Proceedings. Florianópolis: GEDN/UFSC, 2004. p. 554-564.

MARENGO, J. A.; CHOU, S. C.; KAY, G.; ALVES, L. M.; PESQUERO, J. F.; SOARES, W. R.; SANTOS, D. C.; LYRA, A. A.; SUEIRO, G.; BETTS, R.; CHAGAS, D. J.; GOMES, J. L.; BUSTAMANTE, J. F.; TAVARES, P. Development of regional future climate change scenarios in South America using the Eta CPTEC/HadCM3 climate change projections: climatology and regional analyses for the Amazon, Sao Francisco and the Paranà River basins. Climate Dynamics, v. 38, p. 1829-1848. 2011.

MEYER, L.D. AND MANNERING, J.W. Minimum tillage for corn: Its effects on infiltration and erosion. Agricultural Engeneering. 42,2:72-75,86, 1961

MCISAAC, G.F., MITCHELL, J.K., HIRSCHI, M.C. Nutrients in runoff and eroded sediment from tillage systems in Illinois. Summer meeting American Society of Agricultural Engineers. Paper 872066,1987.

MELO, D.; PEREIRA, J. O.; NÓBREGA, L. H. P.; OLIVEIRA, M. C.; MARCHETTI, I.; KEMPSKI, L. A. Physical and structural characteristics of a red Iatosol under no-tillage and minimum tillage systems after four and eight years of no-tillage. Engenharia na Agricultura, Viçosa, MG, v.15, n.3, 228237, 2007.

MENDONÇA, J.C. SOUSA, E.F.; BERNARDO, S.; DIAS, G.P.; GRIPPA, S. Comparison between methods for estimating reference evapotranspiration (ETo) in the Norte Fluminense region, RJ. Revista Brasileira de Engenharia Agricola e Ambiental, v. 7, n. 2, p. 275-279, 2003.

MESINGER, F. A blocking technique for representation of mountains in atmospheric models. Rivista di Meteorologia Aeronautica 44(1-4): 195-202. 1984.

MINISTRY OF AGRICULTURE, LIVESTOCK AND SUPPLY (MAPA). Crop and Livestock Integration, Producer's Primer. Brasilia, 2007.

NOBRE, C. A.; ASSAD, E. D. O.; Global warming and the impact on the Amazon and Brazilian agriculture. INPE e Print, Sao José dos Campos, v.1, 2005.

UNITED NATIONS ORGANIZATION (UN). Available at:

<http://www.onu.org.br/index.php?s=POPULA%C3%87%C3%83O+MUNDIALex=
-1003ey=-165> Accessed: Feb. 2013.

ORTEGA-FARIAS, S. Integral system for water management. In: Modernización de Riegos y Uso de Tecnologias de Información (CYTED-Riegos Network, La Paz, Bolivia, Sept. 2007). Proceedings... La Paz, Bolivia CD-ROM. 2007.

PAZ, V.P.S.; Teodoro, R.E.F.; Mendonça, F.C. Recursos Hidricos, agricultura irrigada e meio ambiente. Revista Brasileira Engenharia Agricola e ambiental, Campina Grande. V.4, n.3, p.465-473, 2000.

PEREIRA, L.S.; VAN DEN BROEK, B.; KABAT, P.; ALLEN, R.G. (eds), CropWater Simulation Models in Practice. Wageningen Pers, Wageningen. USA. 1995.

PEREIRA, L.S., Perrier, A., Allen, R.G., Alves, I., Evapotranspiration: concepts and future trends. J. Irrig. Drain. ASCE 125 (2), 45-51, 1999.

PEREIRA, A. R., Angelocci, L. R., Sentelhas, P. C., Agrometeorology, Fundamentals and Practical Applications. Guaiba/RS - Brazil, 2002.

PEREIRA, LS, Oweis T, Zairi A. Irrigation management under water scarcity. Agric. Water Manage. 57: 175-206, 2002.

PEREIRA, L.S. Indicators of water use. In: AF Cirelli and E.M. Abraham (Eds.). Uso y gestion Del agua en tierras secas. v. XI El Agua en Iberoamérica. CYTED Area IV. 13: 207-214p. 2003.

PEREIRA, L.S. Water Needs and Irrigation Methods. Pulb. Europa-América, Lisbon, 2004, 313 p.

PEREIRA, L. S. Necessidades de Agua e Programaçao da Rega: Modelaçao, Avanços e Tendências, Lisbon, 2007, 25 p.

PESQUERO, J. F. Moisture balance in the monsoon system region of South America in future climate scenarios (2071-2100) using the Eta Model: a modeling study. PhD thesis in Meteorology. National Institute for Space Research. Sao José dos Campos, SP. 242 p. 2009.

PESQUERO J. F, CHOU, S. C., NOBRE C. A., MARENGO, J. A. Climate downscaling over South America for 1961-1970 using the Eta model. Theor Appl Climatol. doi:10.1007/s00704-009-0123-z. 2010.

UNDP Brazil - United Nations Development Program. Available at: < http://www.pnud.org.br/rdh/> Accessed on: 17 Sep. 2011.

PRAPEM - Studies of Water Resources Management Instruments for the State of Santa Catarina and Support for their Implementation. Government of the State of Santa Catarina, 2006.

RADIN, B.; SANTOS, A. O.; BERGAMASCHI, H.; ROSA, L. M. G.; BERGONCI, J. I. Estimation of maize crop evapotranspiration by the modified Penmann-Monteith method. Revista Brasileira de Agrometeorologia, Santa Maria, v.8, n.2, p. 185-191, 2000.

RIBEIRO, R.S.F. Computerized model for determining irrigation schedules. Master's dissertation. Fortaleza/CE. 1992, 89p.

RITCHIE, S.W; HANWAY, J.J and BENSON, G.O. How the corn plant develops. Goiânia: POTAFOS, 2003. 20p (Informaçôes agronômicas, 103).

RODRIGUEZ, F. A., FERNANDES, C., CASTILHA, H. R., VALÉRIO, M. A. Irrigation in Brazil, Situation and Guidelines - MINISTÉRIO DA INTEGRAÇÂO NACIONAL- Brasilia/ Brazil - May 2008

SANTA CATARINA, Government of the State of Santa Catarina - Overview of Water Resources in Santa Catarina. Available at: <http://www.aguas.sc.gov.br> Accessed on; 01 Sept. 2011.

SANTA CATARINA, State Secretariat for Sustainable Development. Directorate of Water Resources (Execution: UNISUL). Diagnosis of water resources and organization of agents in the Tubarao river basin. Florianópolis, 1998.

SEVERO, L. D. Case studies of heavy rainfall in the state of Santa Catarina. Master Science Dissertation, INPE, Sao José dos Campos, Brazil. 1994.

SIAGAS. Groundwater Information System. State of Santa Catarina. Available at: < http://siagas.cprm.gov.br/wellshow/indice.asp>. Accessed on: May 2006.

SILVA, Gilson Carlos. Quantitative analysis of extreme precipitation events in the eastern and northern regions of Santa Catarina and climate projections according to regional climate modeling. Pelotas-RS, 2011. 105f. Dissertation (Master of Science). Postgraduate Program in Meteorology. Federal University of Pelotas, 2011.

SILVEIRA, G.M. O preparo do solo: implementos corretos. Globo, Rio de Janeiro, 243 p. 1988.

SMITH, M. Report on the expert consultations on revision of FAO methodologies for crop water requirements.Rome: FAO, 1991. 45 p. Available at:

<http://www.fao.org/nr/water/docs/Revised-FAO-Methodology-CropWaterRequirements .pdf>. Accessed on: June 11, 2013.

STONE, L. F.; MOREIRA, J. A. A.; Effects of tillage systems on *water* use and bean productivity. Pesquisa agropecuária brasileira, v. 35, n° 4.. p.835-841. Brasilia, 2000.

TESTA, Vilson Marcos et al., The sustainable development of western Santa Catarina: a proposal for discussion. Florianopolis: EPAGRI, 1996, 247 p.

TEDESCO, M. J.; GIANELLO, C.; ANGHINONI, I.; BISSANI, C. A.; CAMARGO, F. A. O.; WIETHOLTER, S. (Ed.). Fertilization and liming manual for the states of Rio Grande do Sul and Santa Catarina. Porto Alegre: Brazilian Society of Soil Science - Soil Chemistry and Fertility Commission, 2004. 400 p.

TEIXEIRA, José Luis; PEREIRA, Luis Santos. ISAREG - An irrigation scheduling model. ICID: Bulletin 41, v.2. 1992. p.29-48.

TIMM, L. C. and REICHARDT, K., Soil, Plant and Atmosphere, concepts, processes and applications - v.2. Sao Paulo, Brazil, 2012.

TORRECILLAS, A.; CONEJERO, W.; ORTUNO, M.F.; GARCIA-ORELLANA Y.,

NICOLAS, E.; ALARCÓN, J.J. Utilización de las medidas de las variaciones del diametro del tronco para la programación Del riego en melocotonero temprano. In: Modernización de Riegos y Uso de Tecnologias de Información (Red CYTED-Riegos, La Paz, Bolivia, Sept. 2007). Proceedings... La Paz, Bolivia CD-ROM. 2007.

ToNIN, G.A.; CARVALHo, N.M.; KRoNKA, S.N.; FERRAUDo, A.S. Influence of cultivar and vigor on the germination performance of maize seeds under conditions of water stress. Revista Brasileira de Sementes, Brasilia, v.22, n.1, p.276-279, 2000.

UNFPA, United Nations Population Fund. Available at:

<http://www.unfpa.org/public/> Accessed on: 14 Feb. 2012.

VAN LIER, Q. de J. Fisica do Solo - Brazilian Society of Soil Science. Viçosa, Minas Gerais, 2010.

VEIGA, M. Routine methodologies for soil physical analysis. EPAGRI/FAPESC Doc. 2011. 62p

VIANA, J.M. Determination of the irrigation schedule for crops in the Curu-Paraipaba irrigated perimeter, using a computer model (Cropwat v.5.7, 1991- FAO). Master's dissertation. Fortaleza-Ce. 1997, 122p.

VICTORIA, F. R. B.; Irregular rainfall distribution and the crop cycle. In: I Corn and Bean Technical Meeting of Santa Catarina. Chapecó, Epagri, 1998. p.59- 62.

VICTORIA, F. R. B., TEIXEIRA, José Luis. Integrated tools to mitigate effects of climatic adversities. In: Inter-Regional Conference on Environment-Water, Proceedings. ICID-ABID 4th : 2001: Fortaleza, Brazil, 2001. p.46-55.

VIEIRA, M.J. Physical behavior of the soil in no-till farming. In: FANCELLI, A.L.; TORRADO, P.V.; MACHADO, J. (Coords.). Update on no-till farming. Campinas: Fund. Cargill, 1985. p.163- -179.

ZHAO, Q., BLACK, T. L., BALDWIN, M. E. Implementation of the cloud prediction scheme in the Eta model at NCEP. Weather Forecast 12:697-712. 1997.

Printed by Books on Demand GmbH, Norderstedt / Germany